Anonymous

Osteopayth;

The new science of healing

Anonymous

Osteopayth;
The new science of healing

ISBN/EAN: 9783337714871

Printed in Europe, USA, Canada, Australia, Japan

Cover: Foto ©berggeist007 / pixelio.de

More available books at **www.hansebooks.com**

OSTEOPATHY

The New Science of Healing.

HUDSON-KIMBERLY PUBLISHING CO.
KANSAS CITY, MO.

INDEX.

ERRATA.

On pages 96, 97, and 100, for "general treatment (page 32)" read "general treatment (page 93)."

EXPLANATORY.

Aponeurosis—A fibrous expansion of a tendon.

Conjunctiva—Mucous membrane of the eye.

Bifurcation—Dividing into two branches.

Poupart's Ligament—Ligament in upper part of thigh.

Areola—A ring-like discoloration.

Arteriole—A small artery.

Radicle—A rootlet.

Sterno-clavicular Articulation—Articulation of the clavicle (collar-bone) and sternum (breast-bone).

Aperients—Substances having power to open passages.

Protoplasm—Primitive organic cell matter.

Urea—Chief solid constituent of the urine.

Uric Acid—Acid normally found in urine.

Bile—Yellow bitter liquid secreted by the liver.

Ileocecal Valve—Valve between the ileum and cæcum.

Crest of the Ileum—Upper free margin of the ileum.

Diaphragm—Muscular wall between the thorax and abdomen.

Tuberculosis—Infectious disease due to specific bacillus.

Hemiplegia—Paralysis of one side of the body.

Paralgia—Disordered sense of pain in a part.

Encephalon—The brain.

Foramina—Openings in the bones for passage of vessels and nerves.

Esophagus—Canal leading from the larynx to the stomach.

Chyme—Food that has undergone gastric, but not intestinal, digestion.

Pharynx—Musculo-membraneous sac behind the mouth.

Larynx—Upper part of windpipe.

Pyloric Orifice—Opening in pyloric end of stomach.

Duodenum—First part of the small intestine.

Thyroid Cartilage—Largest laryngeal cartilage.

Tendon—White fibrous tissue, attachment of the muscles.

Peripheral Nerves—Peripheral, pertaining to the circumference or boundary line.

Ligament—A band of fibrous tissue binding parts together.

Cartilaginous—Pertaining to cartilage—viz., gristle.

Pleura—Serous membrane enveloping the lungs.

PREFACE.

By an examination of the literature of the world, it will be found that the subject of Osteopathy has never yet been placed before the public in book form. Its fundamental principles were discovered by Dr. Andrew T. Still, and a class established by him, of which the author is a graduate.

Knowing that Osteopathy is destined to revolutionize the medical world, and realizing that even a limited knowledge of this most wonderful science would save worlds of suffering and thousands of human lives, as a duty we owe to humanity we have thrust aside the vail of mystery in which it has been shrouded and place it before the thinking world in all its grandeur, simplicity, and truth.

As this book is intended to reach the masses, we will endeavor to avoid as much as possible anatomical words and medical phrases, which too long have been used to confuse and mystify the people.

Dr. Elmer D. Barber.

The author was born in Oneida, N. Y., May 17, 1858, and when a mere boy arrived at the conclusion that the age of miracles was past, and that all results could be traced to a reasonable cause.

While in Jersey City, N. J., we met a gentleman who, without the use of drugs or surgical instruments, by manipulations which he could not explain, was curing hundreds of people in a public hall.

Then came Paul Castor, whose cures were equally marvelous and likewise inexplicable. We visited faith doctors and spiritualistic mediums and witnessed their results, but found the principle on which they worked shrouded in mystery.

We next heard of Dr. Andrew T. Still, who was effecting cure after cure in a marvelous manner, upon (as he claimed) scientific principles. We visited the old doctor, were convinced that he had discovered the fundamental principles on which were based the results accidentally reached by others, and entered his School of Osteopathy, from which we graduated March 2, 1895, with an average grade of 99 in Anatomy and Physiology and 100 in Osteopathy.

While it is our desire to give Dr. Still credit for the new science which he discovered, we must differ with him as to the true cause of the results reached by the Osteopath. While the good Doctor believes that nearly all diseases are caused by dislocated bones, nearly always finding them and thereby winning for himself the name of "Bone Doctor," in our practice we never find a great number of dislocations and by the same manipulation effect the same cures as Dr. Still. If a bone is really dislocated and has been in that condition for years, the dislocation can not be reduced; but if the muscles are contracted, causing a stiff joint or depressing the ribs, they can be quickly relieved by manipulation, and the patient is easily led to believe that the bone was dislocated. While we do not doubt for an instant that our classmates are sincere in their belief that in catarrh, sore eyes, deafness, and other disease of the head, the atlas is dislocated, and that they cure these diseases by setting the atlas, we believe that twisting, pulling, and stretching the neck in a vain attempt to move the atlas stretches the muscles, thereby freeing the circulation and permitting Nature to assert herself. Be they right or wrong, our readers can cure any acute disease in the head, almost instantly, by gently pulling on the head and rotating it in all directions; and any chronic complaint, except cancer, total deafness, or total blindness, by a continuation of the same method. We all agree upon the one great point that man is a machine, and that Dr. A. T. Still has discovered centers upon which a pressure of the hand will cause the heart to slow or quicken its action, from which we can regulate the action of the stomach, bowels, liver, pancreas, kidneys, and the diaphragm. The thousands of people snatched from the grave by an application of these never-failing principles are proof postive that at last the keynote has been struck; that at last a man is found and a school established that can explain intelligently why certain manipulations produce certain results we all agree.

Viewing the brain, the cerebro-spinal cord, and the nerves as an immense telegraph system (the brain acting not only as a great dynamo, generating the forces which control and move the body, but as headquarters, receiving and sending messages to all parts of the body; the slender nerves passing through, under, over, and between the hundreds of bones, muscles, arteries, veins, ligaments, and various organs), can you wonder that the wires are sometimes down, that the communication is occasionally cut off between headquarters and some important office, or that paralysis is the result? Do you wonder that occasionally the wires are crossed, and that the message (possibly to the bowels, to discharge their load) is received by the kidneys, which promptly obey the order? The bowels having failed to respond to orders from headquarters, a second message goes over the wires, and again the kidneys answer the summons; the result is kidney disease and constipation. Will you pour poisonous drugs into the unoffending stomach, which has never failed to obey an order received, or would it be advisable to try and fix the wires? While we cannot go directly to the nerves at fault, we can, by manipulations, which will be fully described under their proper head, stretch the contracted muscle that is obstructing the current; whereupon, if the case has not become chronic, the bowels will immediately resume their functions and the excited kidneys will cease to act so rapidly. In chronic cases it usually takes Nature from two to six weeks to assert herself. In the nervous system, as in the telegraphic, the current must not be obstructed, or disease and death are the result.

The massage treatment, which we believe is based upon strength and ignorance, effects many remarkable cures by moving the flesh and muscles in all possible directions over the entire body. They unwittingly and unavoidably, if they are very thorough, free the right spot, establishing the cir-cuit, thus permitting Nature to assert herself.

Another very important part in this complicated machine is the systemic, pulmonary, and portal circulation: the arteries, cylindrical vessels, conveying the blood through this network of nerves and muscles to all parts of the body; the veins gathering up and returning it to the never-tiring heart, pumping steadily throughout a lifetime, driving the blood to the most remote part of the system and forcing it to return. Is it to be wondered at that occasionally a muscle contracts, after a hard day's work or exposure to the cold, possibly obstructing some little river of blood on its journey to nourish a given part? Do you wonder that the part in question weakens from lack of nourishment and fails to perform its allotted task? As it is the blood that must convey all substances of nourishment to the different parts, is it a wonder that the medicine never arrives at its destination? Should a large artery be obstructed in a similar manner, would it be surprising if the heart, working against heavy odds, trying to pump the blood past the obstruction, in time felt the pressure? in which case heart disease would be the result. Shall we now convert the long-suffering stomach into a drug store, or, viewing man as a machine, remove the cause? Should the contraction be in the thigh, obstructing the femoral artery, we have cold feet and limbs on one side of the obstruction, and heart disease on the other. If the veins returning the blood are obstructed in the same region, we may have either dropsy, inflammatory rheumatism, erysipelas, eczema, or varicose veins, caused by the stagnant, pent-up blood, on one side, and heart disease on the other.

Having briefly referred to the bones that support the nerves that control, and the blood that supplies, let us dwell for a moment on the muscles that move and propel this wonderful living machine. As the only power muscles have is in contraction, they must be arranged in such a position and so attached to the bone as to pull from any direction in which it may be necessary to move a given part. Receiving as they do not only their orders to act, but their motor power, from that great dynamo the brain, they may be justly compared to so many electric cars. One car may be larger and stronger than another, but, deprived of the current from that slender wire, which of itself is nothing, neither can move from its position. Is this not indeed a delicate and complicated piece of machinery, the nerves and fluids of the body moving unobstructed through the hundreds of rapidly contracting and relaxing muscles? We state most emphatically that the true cause of all disease may be traced to some muscle which has contracted and for some unaccountable reason has failed to relax, thus interfering with all the forces of life. It is by working on these principles which we have briefly sketched that we achieve results bordering on the miraculous; it is by working on these principles that Dr. Still draws patients from the length and breadth of our land; it is by working on these same principles, fully explained and illustrated in the following pages, that any family can attain the same results.

OSTEOPATHY IN A NUTSHELL.

First: Using the arms and limbs as levers, stretching all muscles to which they give attachment and moving the flesh and muscles from side to side the entire length of the limb stretches and softens those muscles, thus permitting a free flow of the fluids and nerve forces to these parts, a stoppage of which means disease in some of its varied forms. One thorough treatment of an arm or leg will often instantly cure and always relieve an acute case of any nature in the extremities, and a very few treatments, administered one each day, will *cure* any acute case. Chronic cases can be usually cured by a continuation of the treatment, every other day, for from two to six weeks, even after all other methods have been tried and failed.

Second: Move and soften, by deep manipulations and by rotating the body as much as possible, all the muscles of the spine, the cerebro-spinal cord being the great trunk from which springs the spinal nerves, it being contained in and protected by the upper three-fourths of the spinal column, which is very flexible, consisting of many separate bones,

between which is placed the elastic intravertebral cartilage.
As the spinal nerves which control the different muscles,
organs, etc., escape from the spinal cord through openings
or foramina in the different sections of the vertebral column.
it will be readily understood that the numerous muscles
which are attached to and move the spine must always be
very soft and elastic; that contraction here means inter-
ference with nerves that may control some distant part and
a consequent partial or complete paralysis of that part, until
by manipulation or accidentally you stretch the muscle at
fault, thus turning on the current from that great dynamo
the brain, and once more your machine moves forward.
What would be your opinion of a motorman, when his car
came to a standstill through lack of motor power, if he poured
medicine on the wheels? It would be just as sensible as
converting the stomach into an apothecary's shop, hoping
thereby to remove an obstruction which was breaking the
current between headquarters and the liver. We find that
there are very few organic troubles whose origin may not be
traced directly to the spine and cured by a thorough treatment of the spinal column continued every second day for
from two to six weeks. In 90 per cent of all cases immediate
relief will be the result of the first treatment.

   Third: Using the head as a lever, move and stretch all
the muscles of the neck. This treatment frees the circulation to the head, an obstruction of which is the true cause
of catarrh, weak eyes, deafness, roaring in the head, dizziness,
and, in fact, almost all disorders of the head. Many acute
cases can be instantly cured, while those that have become
chronic require from two to six weeks.

   Fourth: Bending the patient backward, with the knee
pressing on the back just below the last rib, will instantly
cure any case of looseness of the bowels, from common

diarrhea to bloody flux, and a continuation of the treatment
will cure any case of chronic diarrhea.

Fifth: A nerve-center has been discovered at the base
of the brain, termed vaso-motor, which can be reached by a
pressure on the back of the neck over the upper cervicals. A
pressure at this point continued from three to five minutes
will slow the action of the heart, often reducing the pulse
from 100 to a normal condition in a few minutes' time. It is
from this center that, without the use of drugs, we control
fevers, curing any fever that is curable in one-half the time
that the same work can be done with medicine.

Sixth: In all cases where the general system seems to
be affected, give a general treatment, thus freeing and permitting all the forces of the machine to act.

Seventh: Never treat an acute case oftener than once
in three hours, or a chronic case oftener than once a day.

Eighth: It is never safe to use this treatment during
pregnancy, except in diseases of the head or extremities, and
in those with caution. To draw the arms high and strongly
above the head, at the same instant pressing on the spine
below the last dorsal vertebra, or to flex the limbs strongly
against the chest, during this period, is dangerous in the
extreme.

Ninth: While this treatment will improve the action
and remove the pain in stiff, chronic dislocated joints, the
dislocation can never be reduced. We have seen it tried
and tried it ourselves a great many times, meeting with no
success where there was really a dislocation. There are a
great many cases where the patient is suffering from rheumatism or a similar trouble in which the muscles are contracted,
and he can easily be led to believe that a dislocation does
really exist, and that the operator who simply stretches the
muscles has reduced the imaginary dislocation. This we
believe also to be the case regarding the many dislocated

ribs found by the average "bone doctor." While they may be correct, we have demonstrated the fact, times without number, that drawing the arms high above the head, at the same instant pressing at almost any point with the knee immediately below the scapulas, thus stretching the muscles of the chest and springing the ribs forward, will instantly cure sharp acute pains in the sides or chest and certain cases of heart disease, while a continuation of the same treatment will cure asthma or consumption. It is on this vital point that we differed in class as well as in practice with the other members of our profession. While they trace most effects to dislocated bones, and never fail to effect a cure if it is within the bounds of reason, we effect equally remarkable cures by simply stretching and manipulating the muscles, thus freeing the circulation. While we do not believe it possible that to hide his secrets a "bone doctor" would deceive the public, we believe that in a vain attempt to set the bones in the manner prescribed by Dr. Still the circulation is freed and the patient recovers.

HOW TO APPLY OSTEOPATHY.

First: Secure a pine table, two feet high, two feet wide, and six feet long, over which spread a bed-quilt and at one end place one or two pillows. While an acute case may be treated on a chair, a couch, or on the floor, for a chronic case, which is liable to take several weeks' treatment, it is always advisable to secure a table.

Second: In treating a gentleman it is seldom necessary to remove more than his outer clothing.

Third: A lady must loosen her tight clothing and remove her corset. The principles of Osteopathy as described in this work can be applied successfully through a reasonable amount of clothing, except in cases which will be apparent.

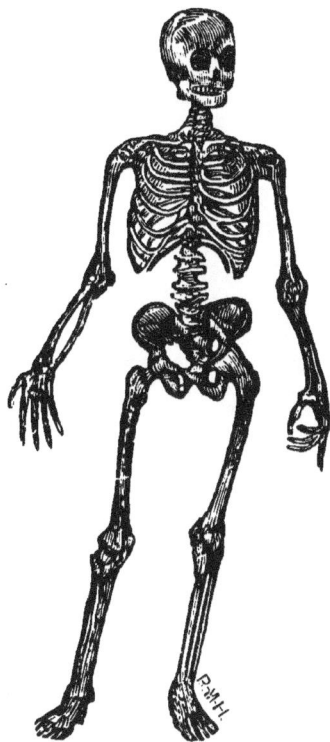

MAN AS A MACHINE.

The entire skeleton in the adult consists of two hundred distinct bones, articulating with each other in perfect harmony. Some are arranged to allow the utmost freedom of motion, others are limited, while others are fixed and immovable. On the bones are many prominences for the attachment of muscles and ligaments and many openings (or foramina) for the entrance of nutrient vessels.

The thorax is a bony cage formed by the ribs, the dorsal vertebræ, and the sternum; it contains the principal organs of respiration and circulation.

Should the muscles of the chest contract, as is often the case, springing the ribs, which are the most elastic bones in the body, lessening the dimensions of the thorax, we have asthma, consumption, or heart disease; while a partial dislocation of the lower ribs, caused by contracting muscles, causes enlargement of the spleen, stomach trouble, and various other diseases which can readily be cured by manipulations.

There are over five hundred muscles in the human body connected with the bones, cartilage, and skin, either directly or through the intervention of fibrous structures called tendons or aponeuroses. Muscles differ much in size; the gastrocnemius forms the chief bulk of the back of the leg, and the fibers of the sartorius are nearly two feet in length, while the stapedius, a small muscle of the internal ear, weighs about a grain, and its fibers are less than two lines in length.

Now, having briefly mentioned the bones and the muscles we will touch upon the arteries that nourish this most interesting and intricate piece of machinery.

CIRCULATION OF THE BLOOD.

The course taken by the blood on its way to the various parts of the body is called the systemic circulation, on account of its having to make repeatedly the circuit of vessels leading to and from the heart.

The arteries are small cylindrical muscular vessels, and might be compared to rivers throwing a branch to each muscle in their course, while the veins gather up and return the venous blood to the heart, where it is pumped through the pulmonary arteries to the lungs.

It will now be readily understood, as the heart is a double pump, driving the blood through the arteries and veins, that the contraction of muscles throwing a pressure on arteries or veins which pass through, under, or between them would certainly affect the heart and necessarily derange the entire system. We trust that our readers will note these points carefully, as we expect to prove that many cases of heart disease, rheumatism, dropsy, neuralgia, tumor, goitre, and cancer are caused by contracted muscles and are readily cured by a system of treatment which removes the cause and gives Nature a chance to act.

To illustrate more fully, and demonstrate the folly of converting the stomach into an apothecary's shop, let us compare the systemic circulation to an irrigating system. Through your fields run innumerable ditches; one is obstructed by a fallen tree, causing the water to back up, seeking some other channel or a weak place in the bank to escape. What is the result? Too much water in one end and too little in the other. Will you go to the reservoir and throw in a little quinine, a little calomel, and a little whisky, or will you remove the cause? Thus it is that when confronted with heart disease you should immediately ascertain if the patient is troubled with cold extremities; such being the case, using the limbs as levers, stretch the muscles as shown in cut 6, page 49, thus freeing the arteries from this undue pressure, permitting the blood to pass down to and warm the extremities, at the same time relieving the heart.

We will now pass to the nerves, which not only control the action of the muscles and various organs, but also control the caliber of the arteries, thus regulating (when not interfered with by slight dislocations of bone or contraction of muscles) with the utmost precision the entire systemic, pulmonary, and portal circulation. The central part of the nervous system or cerebro-spinal axis consists of the spinal cord (medulla spinalis), the bulb (medulla oblongata), and the brain; the spinal cord being the great bond of connection between the brain and the majority of the peripheral nerves. As most of the nerves originate in the spinal cord, and as the cord is in direct communication with and might be considered part of the brain, it will be readily understood that a pressure on any of these nerves, interrupting communication between the brain and some distant part, will cause paralysis of the part controlled by the nerve involved.

The digestive organs, liver, pancreas, kidneys, and even the action of the heart, can be regulated by a slight pressure of the hand on certain nerve-centers in the spine, which will be discussed more fully in another chapter.

While we have touched but briefly on the anatomy and physiology of the human body, and it would be most interesting and instructive to read up in Gray and Landois, we trust we have proven to our readers that man is a machine, and laid the foundation for a thorough understanding of our method of treating diseases by manipulation and without the use of drugs or surgical instruments.

ASTHMA.

Symptoms.

Recurrent and temporary difficulty in breathing, accompanied by a wheezing sound and a sense of constriction in the throat, with cough and expectoration. Authors distinguish two varieties: dry convulsive or nervous, and humid or common. In the first variety the attacks are sudden and violent and of short duration, the sense of constriction is hard, dry, and spasmodic, the cough slight, and expectoration scanty and only appearing toward the end of the paroxysm. In the second variety the paroxysm is gradual and protracted, the constriction heavy, laborious, and humid, the cough violent, and expectoration commences early, and is at first scanty and viscid, but afterward copious, affording great relief. In many cases the attack is in the night, and most frequently an hour or two after midnight.

Cause of Asthma.

Asthma, pronounced incurable by the medical fraternity, can be relieved and in most cases cured by an application of the principles laid down in the following pages.

As we have mentioned before, the thorax is a bony cage, formed by the ribs, dorsal vertebræ, and sternum, containing and protecting the principal organs of circulation and respiration. The ribs are not only very elastic, but, being connected with the sternum by costo-cartilage and with the dorsal vertebræ by ligaments, have limited motion. Thus it will be seen that they are easily affected by accident or contraction of the muscles. In most cases of asthma a slight depression will be noticed over the second, third, and fourth ribs on the left side, about two inches to the left of the median line, while the cartilaginous portion of the corresponding ribs on the right

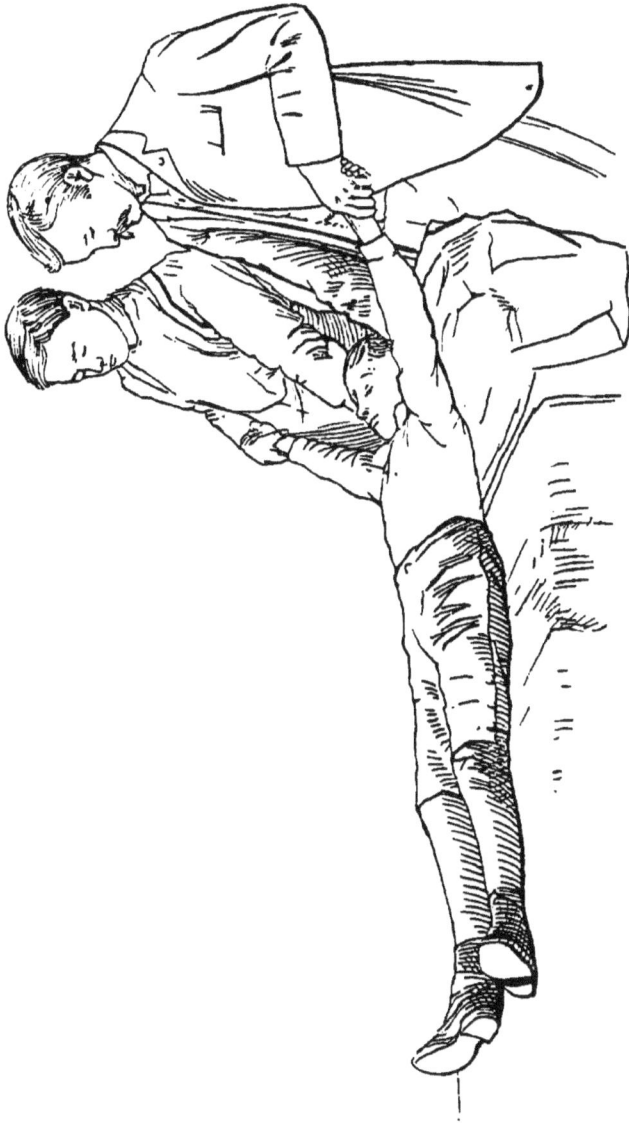

Cut 1.

side will be found elevated; occasionally this will be reversed, but in either case it is proof positive that the framework which is supposed to protect the vital machinery of life is out of gear. Now, having correctly diagnosed this case, having proven that the lungs or bronchial tubes are not at fault, shall we act on the principle that the human system is a machine, and proceed to remove the cause?

Let us suppose you were caught in a cyclone and a heavy timber pins you to the earth. Will you then beg for medicine, or ask some friend to remove the weight that bears you down? If you are suffering from asthma, dear reader, each faint cough, each gasping breath is a prayer from the imprisoned organs within for some one to raise the ribs and expand the chest.

The great Creator, in His infinite wisdom, has arranged for just such an emergency as this by preparing a system of levers, one of which we will now use in raising the ribs, stretching the intercostal muscles, and expanding the chest.

The pectoralis major, a large muscle which covers the entire front of the chest, attaching to the sternal half of the clavicle (collar-bone), the six or seven upper ribs, and the cartilages of all the true ribs, is inserted by a flat tendon into the external bicipital ridge of the humerus about two or three inches below the shoulder-joint. If you will raise your arm high above the head, you will feel all the upper ribs move, thus proving that our theory is correct.

Asthma Treatment.

Place the patient on the back, with a pillow under the head. Two assistants at the head of the table; one places his right, the other his left hand under the patient's shoulders on the angle of the second rib, half way between the scapula (shoulder-blade) and spine and one inch above the scapula. With the disengaged hands take the patient's wrists, and, slowly drawing the arms upward high above the head (see cut 1, page 25), pull steadily and strongly for a moment; at the same time with the fingers press steadily on the angle of the ribs. Lower the arms slowly, the elbows passing below and to the sides of the table. Move the fingers down the spine one inch, to the angle of the next rib, and draw up the arms as before, and repeat until you have raised the four or five upper ribs. It will be also observed that this operation stretches the intercostal muscles.

The patient will now be seated upon a stool. The operator places his knee between the shoulders, grasps the patient's wrists and raises the arms slowly but strongly high above the head, pressing hard with the knee and lowering the arms with a backward motion (see cut 2, page 29).

Instant relief is often felt after the first treatment, and a continuation of the treatment seldom fails to effect a cure. Coughs, colds on the lungs, short, difficult breathing, and pleurisy never fail to respond quickly to our asthma treatment. Of the numerous cases treated by us in this manner 90 per cent have been cured and all benefited.

If every family who reads these pages would make a cheap pine table two feet high, two feet wide, and six feet long, and make a practice of treating each other as directed, they would soon overcome the awkwardness at the first experience, and be able to relieve themselves and friends of innumerable ailments.

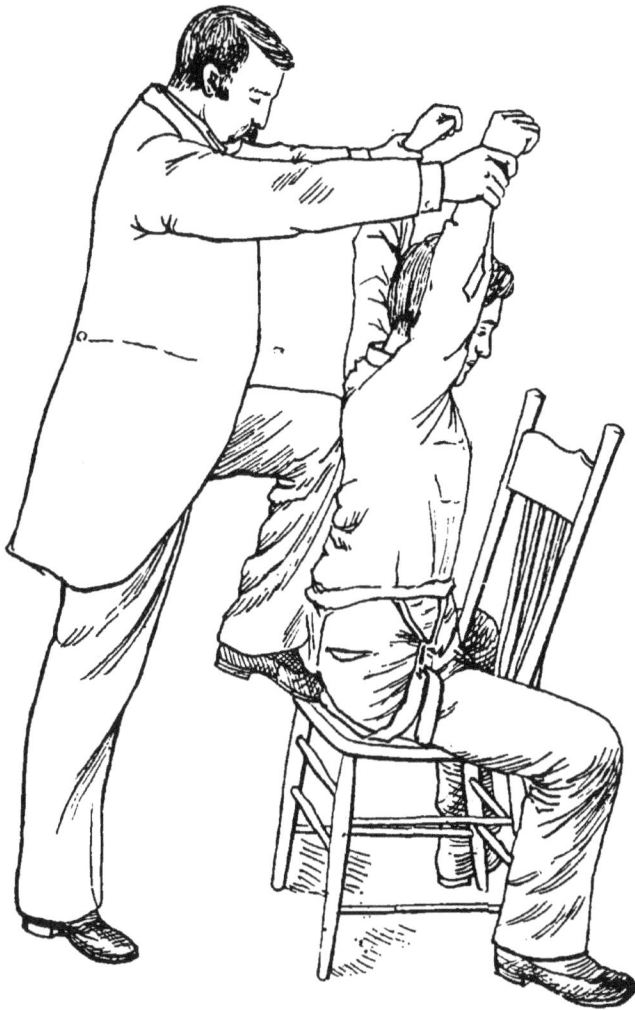

Cut 2.

CONSUMPTION.

Symptoms.

The special symptoms are a short and tickling cough; the pain in the chest is slight, and there is either a sense of tenderness or weight experienced at the upper part of the lungs; the breathing is habitually short, and a full inspiration is impracticable, the attempt increasing the sense of weight and soreness or aggravating the cough; the expectorations are generally scanty and small in quantity in the early stages, and in many cases are very trifling throughout; the matter expectorated is watery and whey-like, sometimes tinged with blood, and as the disease progresses thick, tenacious, curdy, or cheesey particles are excreted. As the functional powers of the lungs become impaired the pulse becomes frequent and feeble, the breathing grows shorter; irregular chills come on, succeeded by some degree of feverish heat, and in the last stages night-sweats, diarrhea, swelling of the limbs, etc., denote the rapidly approaching fatal termination. The local condition of the part diseased is one of engorgement, and its secretions are changed from a healthy to a morbid condition.

It is a well-known fact that cold will contract not only iron and steel, but the muscles of the human body. To prove our theory is correct, allow a cold draught of air to strike the neck for a short time, and possibly the next morning you have a stiff neck. Why is it that the head does not turn freely on its axis? Because the muscles that were exposed have contracted and are a little too short. Acting on these principles, we trace consumption to the contracted muscles of the chest, which are forcing the elastic ribs down upon the pleura and lungs. The old idea is that as the lungs decay the ribs settle. How absurd to imagine that the soft, spongy lungs support the chest! As well say that a house full of sponges would hold up the roof.

We have established the fact, beyond the shadow of a doubt, that it is the steady pressure of the contracting muscles that causes this dread disease, and experience has taught us that until tuberculosis sets in it can be cured.

Consumption Treatment.

Give our asthma treatment, working down as low as the eighth rib, as shown in cut 1, also giving the treatment as shown in cut 2, using great care to exert as much strength as the patient can stand without much inconvenience. In treating consumption, besides our regular asthma and consumption treatment, we usually give a general treatment of the spine, to tone up the system and equalize the circulation.

General Treatment of Spine.

Place the patient on the right side, facing the operator, his left arm flexed, the elbow resting on the right arm of the operator, pressing against the humerus, thus making a lever of the patient's arm to stretch the muscles of the shoulder and scapula (shoulder-blade). The patient must allow his muscles to relax as much as possible. The operator will now place his hands in the position shown in cut 3, page 33. With the finger-ends close to the spine, pressing quite hard, using the arm as a lever, with a circular motion move the muscles under the hand toward the head.

Do not let the hands slip on the spine, as that would be simply rubbing. Our object is to loosen and stretch the muscles, thus freeing the vital forces of life from any obstruction and equalizing the circulation. After each upward motion, move the hands down one inch, keeping close to the spine and working deep the entire length of the spinal column. The left side will now be treated in a like manner. It should not take over fifteen minutes to give the entire treatment, and the effect on the patient is simply wonderful.

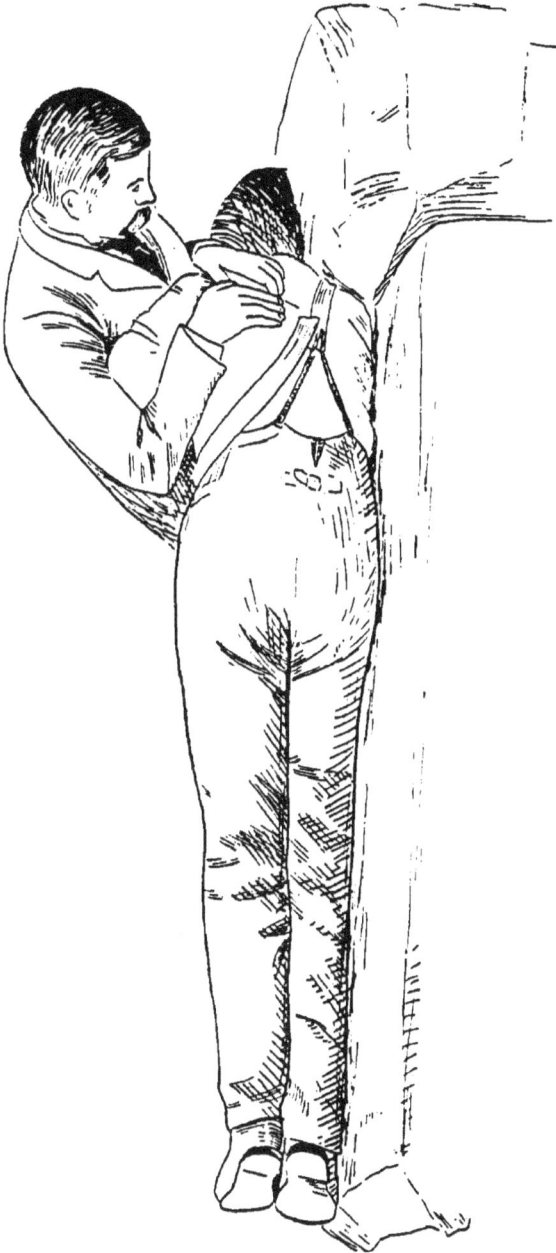

Cut 3.

All manipulation must be slow, careful, and strong. The patient can usually be relied upon to caution the operator if too much strength is being used. A thorough treatment every other day is often enough, as Nature must be given a chance to do her part. Light cases of lung trouble can be cured in two weeks by this treatment, and the most stubborn in from six to ten.

BRONCHITIS.

Symptoms.

This disease is attended with more or less suffocation and spasmodic respiration. The disease commences with more or less cough, irritation about the throat, sense of tightness in the chest, and shortness of breath, which do not for a considerable time attract attention. The first difficulty generally noticed as of importance is a sense of roughness, with frequent attempts to clear the throat, accompanied with or followed by titillation of the larynx, exciting a dry, hard cough. These are, after a longer or shorter period, succeeded by some hoarseness of voice, with a sense of tightness across the chest, and sometimes a siight pain or diffused soreness upon coughing or inflating the lungs fully by a prolonged and deep inspiration. All the causes of consumption under modified circumstances produce this form of pulmonary disease.

Treatment.

Bronchitis being a modified form of consumption, our asthma and consumption treatment (page 27) is especially applicable, giving immediate relief, and in from two to six weeks effecting a permanent cure. Treatment is to be given every other day.

DIPHTHERIA.

Symptoms.

Diphtheria is divisible into two forms, simple and malignant. In the simple variety, happily the most common, the symptoms are at first so mild as to excite little complaint beyond a slight difficulty of swallowing or pain in the throat, burning skin, pains in the limbs, etc.

Malignant diphtheria is ushered in with severe fever, rigors, vomiting or purging, sudden great prostration and restlessness,anxious countenance,etc.,pointing to some overwhelming disease under which the system is laboring. The skin is hot, the face flushed, the throat sore, and the mucous membrane of the throat bright red; the tonsils are swollen, and gray or white patches of deposit appear on them, small at first, but gradually enlarging, so that one patch merges into another, forming a false membrane in the throat, rendering swallowing and even breathing difficult; in some cases the false membrane has been detached and after extreme effort ejected, presenting nearly an exact mold of the throat. The exudation of diphtheria may be distinguished from a slough by its easily crumbling, by the facility with which it can often be detached, and by the surface thus exposed being red, but not ulcerated. The glands of the neck are always enlarged; sometimes pain is felt in the ear, and there is generally stiffness of the neck.

As the disease progresses, the patient passes into a stupor, and the difficulty of swallowing or breathing increases till the false membrane is ejected, or the patient dies from suffocation, or he sinks from exhaustion similar to that observed in typhoid fever.

Dangerous symptoms are a quick, feeble, or very slow pulse, persistent vomiting, drowsiness, delirium, suppressed urine, and bleeding from the nose. Diphtheria is caused by

Cut 4.

a contraction of the muscles of the neck and thorax, as well as by a contraction of the muscles of respiration, which, interfering with the circulation of the fluids of the body, cause the inflamed condition of the larynx, bronchial tubes, and throat. Diphtheria, even in its most malignant form, succumbs to the following treatment.

Diphtheria Treatment.

1st. Place the patient on the back, with one hand under the chin, the other under the back of the head; pull gently, rotating the head from side to side (cut 4).

2d. Pull slowly and strongly until the body moves, without rotating the head.

3d. With the fingers, beginning under the chin, move all the muscles of the neck from side to side.

4th. Place the finger in patient's mouth and move the muscles of the throat gently; this loosens the membrane, which usually will be immediately expelled.

5th. With one hand draw the arm high above the head; at the same instant, with the fingers of the other between the spine and scapula at its upper border, press firmly on the angle of the rib, lower the arm with a backward motion, move the fingers one inch down the spine, draw up the arm, and repeat until the lower border of the scapula is reached. Treat the other side in a similar manner.

6th. Place the arms around patient's body, the fingers meeting at the spine immediately below the last ribs, and, while pressing with the fingers on each side of the spine, raise the patient's body slowly and gently until only the hips and shoulders rest on the bed; this should be repeated two or three times, moving the hands two or three inches each time toward the head; it will instantly stop all purging and vomiting.

7th. Place one hand on each side of the neck, the fingers

almost meeting below the occipital bone (see cut 5); press gently for two or three minutes. It is here you reach nerves that control the caliber of the arteries, thus slowing the action of the heart.

Diphtheria in its most malignant form has never, in our experience, failed to yield readily to this treatment, instant relief being experienced and a complete cure effected in a very few days. Treatment should be given every six hours, and the vaso-motor center (cut 5) may be held at any time, as it always gives relief.

MEMBRANEOUS CROUP.

Symptoms.

Fretfulness, feverishness, cold in the head, and slight hoarseness, increasing towards evening and in the early night. Sometimes, however, without a single warning symptom, the child startles us in the night with a hoarse, ringing cough, which cannot be so described as to be recognized, but which no one who has ever heard it can fail to know again. There is a sense of suffocation, a hurried, hoarse, and hissing breathing, as if the air were drawn into the lungs and expelled through too small an opening in some instrument, which is the fact, for such an instrument is the accumulated phlegm in the larynx. When there is much fever and inflammation, the tendency to the formation of false membrane is very slight; whereas, in cases that seem mild at the beginning the disease often passes to the membraneous stage unsuspected.

Treatment.

Our diphtheria treatment (page 39) will give instant relief, and a permanent cure in a few days if the treatment is continued. A treatment should be given upon retiring each night until all symptoms have entirely disappeared.

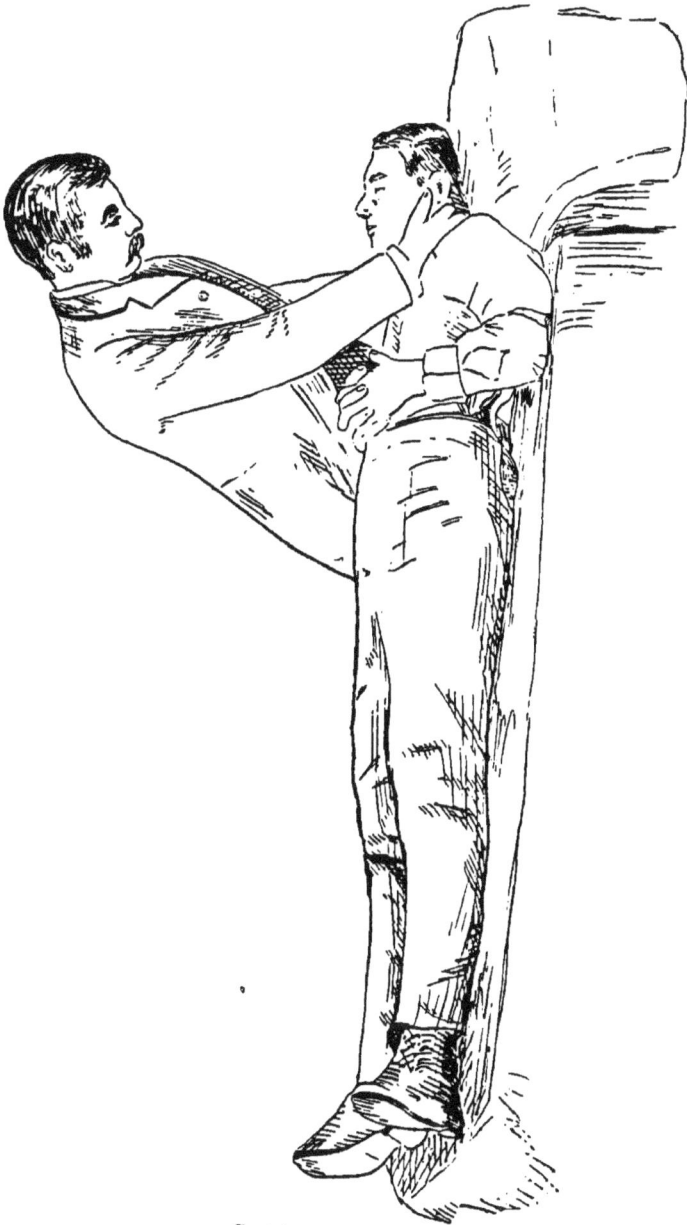

Cut 5.

()

NURSING SORE MOUTH.

Symptoms.

This is an affection from which nursing women occasionally suffer. It consists of inflammation of the lining of the mouth, which is covered with very small ulcers; these cause a stinging and burning sensation, and cheese-like matter exudes from them; a profuse flow of saliva is also frequently present.

Treatment.

1st. Place one hand under the chin, the other under the back of the head (cut 4); pull gently, rotating the head, endeavoring to move and stretch all the muscles of the neck, thus freeing the circulation, which is at fault.

2d. With the fingers, beginning close under the chin, move all the muscles very deep, but gently, the entire length of the front of the neck. This simple treatment in a few days will cure any case of nursing sore mouth. Treatment should be given each day.

PUTRID SORE THROAT.

Symptoms.

This disease generally affects the glands of the throat, while the common quinsy affects the mucous membrane. In putrid sore throat there are also canker sores and ulcers in the throat, together with great debility of the system. In inflammatory sore throat there is always great difficulty of swallowing, whereas in the other these symptoms are not present.

Treatment.

This disease being caused by a contraction of the muscles of the neck, obstructing the blood in the returning veins, which, becoming putrid, works its way to the surface, forming sores and inflammation in the mucous membrane lining the throat, can be easily and quickly cured by our nursing sore mouth treatment (page 43). Treatment should be given each day.

CEREBRO-SPINAL MENINGITIS.

Symptoms.

Acute or excruciating pains in the head, throbbing of the temporal and carotid arteries, flushed face, eyes injected and brilliantly reddish, contracted pupils, and a wild expression of countenance characterize the disease when fully formed. These symptoms are preceded by various cerebral and febrile disturbances—sometimes violent delirium, at other times nausea and vomiting or general convulsions. The bowels are usually extremely costive; there is also great intolerance of light and sound and incessant watchfulness; the skin is dry and hot, the pulse hard and quick, the tongue dry and covered with a white fur, and there is intense thirst.

Cause of Cerebro-Spinal Meningitis.

Cerebro-spinal meningitis, which carries thousands to an untimely grave each year, yields readily to this method of treatment. So simple is the treatment, and so plainly will we place it before the public, that a child can treat it understandingly and successfully, thus robbing that dread malady of at least half its terrors. It was the loss of four of his children by this disease that turned the thoughts of Andrew T. Still, that deep thinker and wonderful reasoner, toward the fundamental principles of this science, which is certain to revolutionize the medical world. After years of study, the good doctor discovered that the muscles which attach to and bind together (in connection with the ligaments) the spinal vertebræ were contracted, thus throwing an undue pressure on the intervertebral substance and interfering not only with the nerve-supply, but with the circulation of the cord itself; and, for the first time looking on man as a machine, he applied the only rational remedy.

Treatment.

1st. Place the patient on his back, and, while gently rotating the head from side to side, stretch the muscles of the neck in all directions (cut 4), using increased strength as the patient becomes accustomed to the motion.

2d. With one hand under the patient's chin, the other under the back of his head, pull gently and gradually stronger until you move the body. It is well to treat the entire length of the spine as shown in cut 3, after which hold the vaso-motor center (cut 5). (The vaso-motor is a nervous center which controls the caliber of the arteries, and can be reached by a pressure on the four upper cervical vertebræ; from this point we control even the action of the heart.) Place one hand under each side of the neck (standing at the patient's side) until the fingers nearly meet, covering a space from the head down the neck the breadth of four fingers; press gently, and in a few minutes your patient will be resting easy. Treatment should be given every six hours, and the vaso-motor center may be held at any time to give relief. This, in hundreds of cases, has never been known to fail, even when death seemed very near; and, when taken in time, we would as soon treat a case of cerebro-spinal meningitis as a hard cold.

PALSY (PARALYSIS).

Symptoms.

Palsy is a disease principally affecting the nervous system, characterized by a loss or diminution of motion or feeling or of both in one or more parts of the body. When one entire side of the body from the head downward is affected, it is distinguished among professional men by the name of hemiplegia; if the lower half of the body be attacked by the disease, it is named paraplegia; and when confined to a particular limb or set of muscles, it is called paralysis. Palsy usually comes on with a sudden and immediate loss of motion and sensibility of the parts. It is sometimes preceded by a numbness, coldness, and paleness, and sometimes by slight convulsive twitches. When the head is much affected, the eye and mouth are drawn on one side, the memory and judgment are much impaired, and the speech indistinct; if the extremities are affected, it not only produces a loss of motion and sensibility, but a wasting of the muscles of the affected parts. The attack is usually preceded by some of these symptoms, but occasionally the disease advances more slowly; a finger, hand, or arm, or the muscles of the tongue, of the mouth, or of the eyelids, being first affected.

Cause of Paralysis.

Paralysis in its various forms, while stubborn, can invariably be cured by our method if taken in time, and even in its last stages great good may be done the patient. Osteopathy is the only rational method of treating this disease. It is caused by a pressure on some of the various nerve-centers, or paralysis of any part may be caused by pressure on the nerves which control that part. The brain, cerebro-spinal cord, and nerves may justly be compared to an immense telegraph system, the nerves carrying messages to and from the brain.

That the average reader may have a more correct understanding of the nervous system, of the power which causes the heart to beat, the blood to flow, the lungs to inhale, the alimentary canal to perform its allotted task, and the muscles to act, let us dwell for a moment on the brain, the spinal cord, and the vertebral or spinal column, as it is on this line of thought that we must reach paralysis.

The nervous system is composed—

First: Of a series of large centers of nerve-matter, called collectively the cerebro-spinal center or axis.

Second: Of smaller centers, called ganglia.

Third: Nerves connected either with the cerebro-spinal axis or ganglia; and

Fourth: Of certain modifications of the peripheral terminations of the nerves, forming the organs of external sense.

The cerebro-spinal center consists of two parts, the spinal cord and encephalon or brain; the latter may be subdivided into the cerebrum and cerebellum, the pons varolii and the medulla oblongata.

The spine is a flexible and flexous column, composed of thirty-three separate and distinct bones in the child, and twenty-six in the adult, articulating with each other and the ribs, enclosing and protecting the spinal cord, supporting the head and trunk, and permitting the escape through its numerous foramina of the nerves which control the body. Is not this indeed a grand and wonderful piece of mechanism? So strong, so delicate, so perfect!

It is to this part of the human machinery that we trace half the ills that flesh is heir to. It is here that we find centers on which a simple pressure of the hand will instantly cure cholera morbus, flux, cramp in the stomach, vomiting, etc., and it is here we must search for the cause and cure of paralysis.

How to Make the Examination.

Place the patient on his face and carefully examine the spine; in perfect health the vertebræ are all in line. If you find one of the spinous processes a little out of line, you have discovered the cause. It may be the result of an accident, it may be turned slightly out of its normal position by a contracted muscle; be that as it may, we have here a pressure on the spinal cord, causing partial or complete paralysis.

Paralysis Treatment.

1st. Free the muscles thoroughly (as shown in cut 3) on each side of the spine the entire length, particularly at the seat of trouble.

2d. Let one assistant now take the patient's shoulders, another his feet, and pull steadily, slowly, and strongly, while the third presses the spinous process back in its place. It may take several treatments, but you will gain a little every time and finally succeed. There is absolutely no danger connected with this treatment if ordinary care is used.

3d. Place the patient on his back, and, with one hand under the chin and the other under the back of the head, pull steadily until the body moves. This must not be omitted, as it starts up circulation in the spinal cord, and even reaches the brain.

4th. Raise the arms and ribs as in cut 1, page 25.

5th. Treat the limbs for circulation and to stretch the great sciatic, the largest nerve in the body, measuring three-quarters of an inch in breadth, and which is the continuation

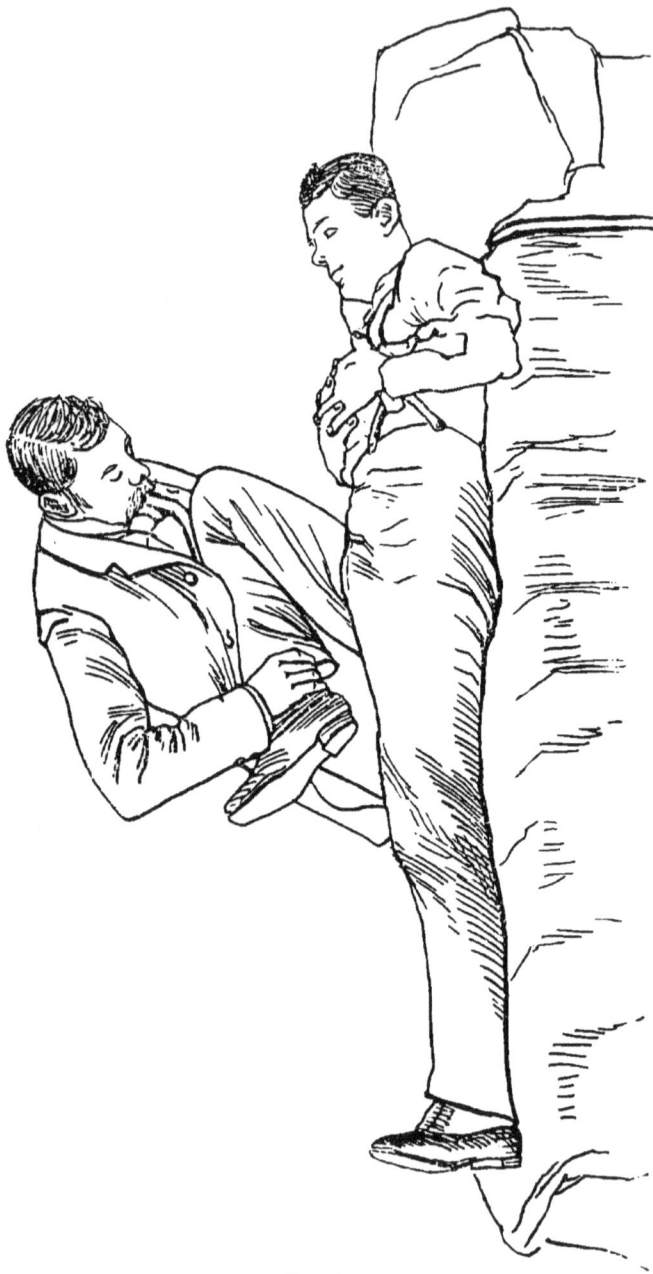

Cut 6.

of the lower part of the sacral plexus; it passes out of the pelvis through the great sacro-sciatic foramen below the pyriformis muscle and descends between the great trochanter and the tuberosity of the ischium, along the back of the thigh, to about its middle third, where it divides into two large branches, the internal and external popliteal nerves. To stretch the sciatic nerve, place the patient on his back, stand at the side of the table, and grasp with the right hand the right ankle, your left hand resting lightly on the patient's knee; now flex the leg slowly against the abdomen as far as possible, using as much strength as the patient can stand (see cut 6). While in this position move the knee three or four times from right to left, without relaxing the pressure; now slowly extend the leg, throwing the knee to the right, the foot to the left, as shown in cut 7. This should be repeated two or three times, and be reversed occasionally, throwing the knee to the left and the foot to the right. Treat the left limb in a similar manner. A treatment should not occupy over fifteen minutes, and should be given every day. This treatment will benefit and usually cure paralysis in any of its varied forms. If no dislocation is found, give the same treatment, as you may have overlooked it, and simply stretching the body will allow a very slight dislocation to slip back and free the cerebro-spinal cord.

Of the numerous cases of paralysis successfully treated by us, we might mention a little girl at Benjamin, northeastern Missouri. She had been a bright, active child until two weeks before she was brought to us for treatment, when it was noticed that she was losing the use of her lower limbs. We explained to the mother the cause of the trouble, a slip in the lumbar vertebræ throwing a pressure on the spinal cord, thus partially cutting off communication between the limbs and the brain. Her old family doctor insisted that the child had worms, and treated her for the same for four weeks, when the little sufferer had entirely lost the use of her lower extremities; he turned her over to us with the remark that he did not understand the case. After four treatments as laid down in this work, the child could walk, and in three weeks was romping with her playmates.

We will refer also to the case of a young lady of Galena, Kansas, suffering from creeping paralysis, or locomotor ataxy. This dread malady is caused by a diseased condition of the posterior column of the spinal cord, and our treatment, stretching and rotating the spine thoroughly, frees the cord and starts the circulation. The young lady in question was not only perfectly helpless, but her digestive organs and kidneys failed to act. In connection with our usual paralysis treatment, we gave the kidney and constipation treatment (see page 69). In a short time the lady could walk with assistance, and in two months was on the high road to perfect health.

ATROPHY (SHRINKING OF MUSCLES).

Atrophy of any part might well be mentioned at this time, it being a form of paralysis. It will be remembered that we mentioned the fact of the nerves controlling the caliber of the arteries, thus regulating the blood-supply. In atrophy we are confronted with a condition in which the nerves controlling the arteries which feed the withered parts are interfered with. The wires are down and the cry of hunger from the starving muscles never reaches the brain.

But even assuming that the brain is apprised of the fact that certain muscles are starving, that they need more blood, its message to the arteries to expand never reaches its destination. If it is the muscles of the leg that are starving, why not flex the leg upon the abdomen (see cut 6, page 50),

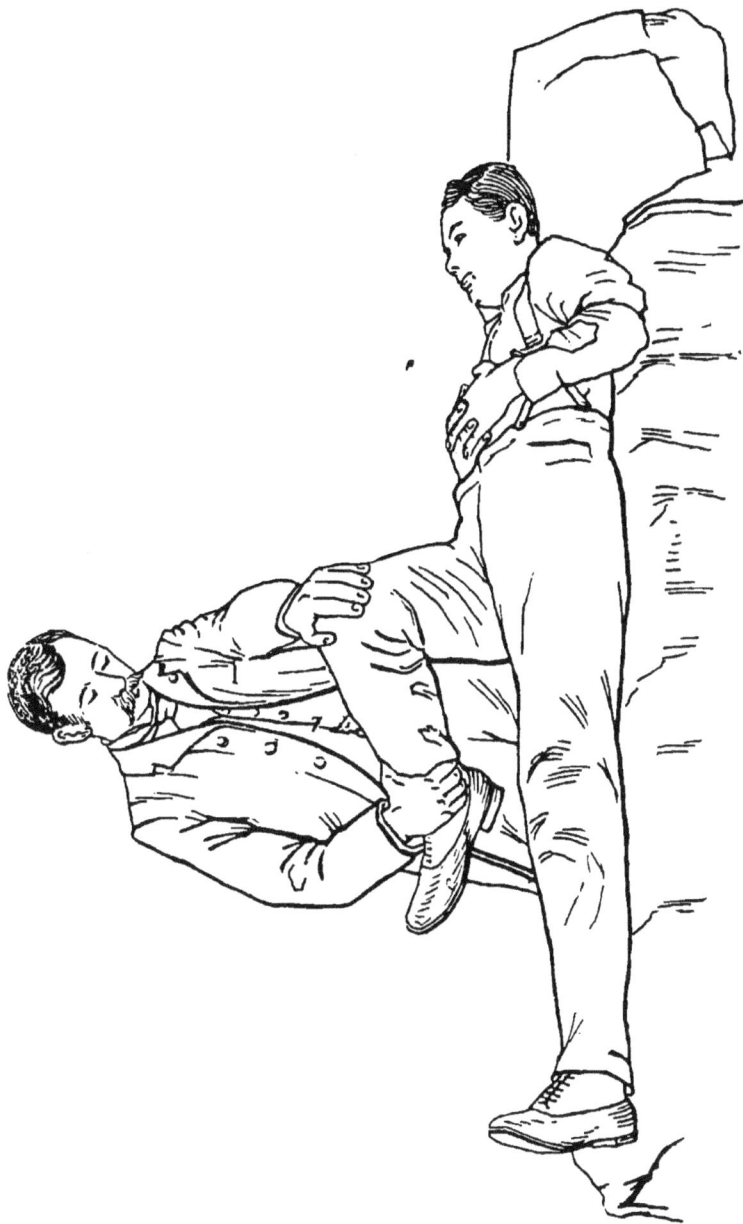

Cut 7.

rotating it inward and outward, thereby stretching the muscles and freeing and stretching the nerves at fault? We have absolutely never known this method to fail to restore shrinking limbs to their normal size. We have cured case after case by this simple, reliable and infallible method, and there is no reason why any of our readers should not be equally successful.

A good general rule in all cases of atrophy of the muscles is to use the affected member as a lever with which to stretch the muscles connecting it to the body in all possible directions, as it is here the obstruction is usually found, and by acting on these principles you will unwittingly stretch the right muscle, thus freeing the nerve and permitting the blood to pass down and nourish the affected parts. It is simply wonderful how quick Nature will respond. If a measurement of the shrunken member is taken, you will know exactly how fast you are progressing; and I venture the assertion that in one month, giving a treatment every other day, you will have gained from one to two inches.

One old gentleman treated by us in Baxter Springs, Kansas, whose leg had been shrunken for years, grew an inch and a half in one month's treatment, measured around the ankle. Another, a young man of the same city, had his arm restored to usefulness in the same remarkable yet simple manner. We would not be understood as saying the limb will gain much in length in the adult; the great improvement will be noticed in size and strength.

DYSENTERY, OR BLOODY FLUX.

Symptoms.

Dysentery usually commences with severe pains in the bowels, with frequent inclination to go to stool; the stools are small in quantity and sometimes mixed with blood.

There is generally a peculiar sensation of bearing down while at stool, as if the whole bowels were falling out, accompanied with considerable pain. Sometimes chills and fever precede the symptoms; in other cases they either accompany them or soon follow, if the attack is at all severe. Griping pains in the abdomen, followed by discharges from the bowels, are the first prominent symptoms.

CHOLERA MORBUS.

Symptoms.

Cholera morbus is a violent purging and vomiting, attended with gripes and a constant desire to go to stool. It comes on suddenly, and is most common in autumn. There is hardly any disease that kills more quickly than this when proper means are not used in due time for removing it. It is generally preceded by heartburn, sour belchings, and flatulence, with pain in the stomach and intestines; to these succeed excessive vomiting and purging of green, yellow, or blackish colored bile, with distention of the stomach and violent griping pains. There is likewise great thirst, with a very quick unequal pulse, and often a fixed acute pain about the region of the navel. As the disease advances, the pulse often sinks so low as to become quite imperceptible; the extremities grow cold or cramped and are often covered with a clammy sweat, the urine is obstructed, and there is palpitation of the heart. Violent hiccoughing, fainting, and convulsions are the signs of approaching death.

Cause of Flux and Cholera Morbus.

Flux and cholera morbus can be cured instantly. Of the hundreds of cases treated by us, we have yet to find the first that did not respond instantly, and we stand ready to wager our reputation that there never was and never will be a case of either of these diseases that cannot be cured by this method of treatment if properly applied. This may justly be considered the grandest discovery of this or any other age. And we beg the medical fraternity throughout our land, who usually look with eyes of skepticism on anything out of the ordinary, to try this one great principle, which is destined to save thousands of lives each year. We trust that each and every one who may chance to read these pages will remember our treatment for flux. Not because it is more reliable than any other great principle laid down in this work, but it is so simple and of such vast importance in times of need, so infallible, and gives such immediate relief.

That the reader may gain a correct understanding of this great principle, we will return to the anatomy, to the machinery of human life. Once more comparing the cerebro-spinal cord, the brain, and the nerves to a telegraphic system, we will trace the cause of flux and the excited condition of the digestive organs directly to the brain. It will be wise, in this connection, as some of our readers may not be very familiar with anatomy and physiology, to sketch briefly the process of digestion.

Food, when taken into the mouth, undergoes two processes, which are inseparable and simultaneous in action, being mastication and insalivation. In the short time occupied by the passage of the food through the œsophagus no special change takes place. In the stomach the food is mixed with the juices of that organ, and is converted into chyme. The chyme begins to leave the stomach through the pyloric orifice soon after gastric digestion has begun, some passing into the duodenum in about half an hour. The materials which resist gastric secretion or are affected very slowly by it are retained many hours in the stomach, and the pylorus may refuse exit to such materials for an indefinite length of time, so that, after causing much uneasiness, they are finally removed by vomiting. Many solid masses escape through the pylorus, however, when it opens to let out the chyme.

The small intestine is a convoluted tube, varying in length from twenty to thirty feet, which gradually diminishes in size from its commencement to its termination. The power which forces the food and chyme through this long convoluted tube is called the peristaltic action, and is controlled by the "main battery," the brain. A wave of contraction passes from the pylorus along the circular fibers so as to look like a broad ring of constriction, progressing slowly downward. The longitudinal fibers at the same time contract so as to shorten the piece of intestine immediately below the ring of constriction, and also causes a certain amount of rolling movement of those loops of intestine which are free enough to move. In flux and cholera morbus this peristaltic action becomes increased to an alarming extent. Food has been taken into the stomach, to remove which a great amount of nerve power is required; and when it is finally expelled, and the current still on, we have a machine running away with itself. We are as yet unable to determine the precise cause of Nature failing to apply her brakes and check the current at the proper moment, but we have succeeded in locating the point on which a slight pressure of the hand will instantly slow up the machine.

The great splanchnic and right pneumogastric nerves form the solar plexus, or great abdominal brain, and control the peristaltic action of the bowels. Now it is obvious that a pressure on these nerves long enough to break the current will check the peristaltic action of the intestines. The pneumogastric has a more extensive distribution than any of the other cranial nerves. Passing through the neck and thorax to the upper part of the abdomen, it is composed of

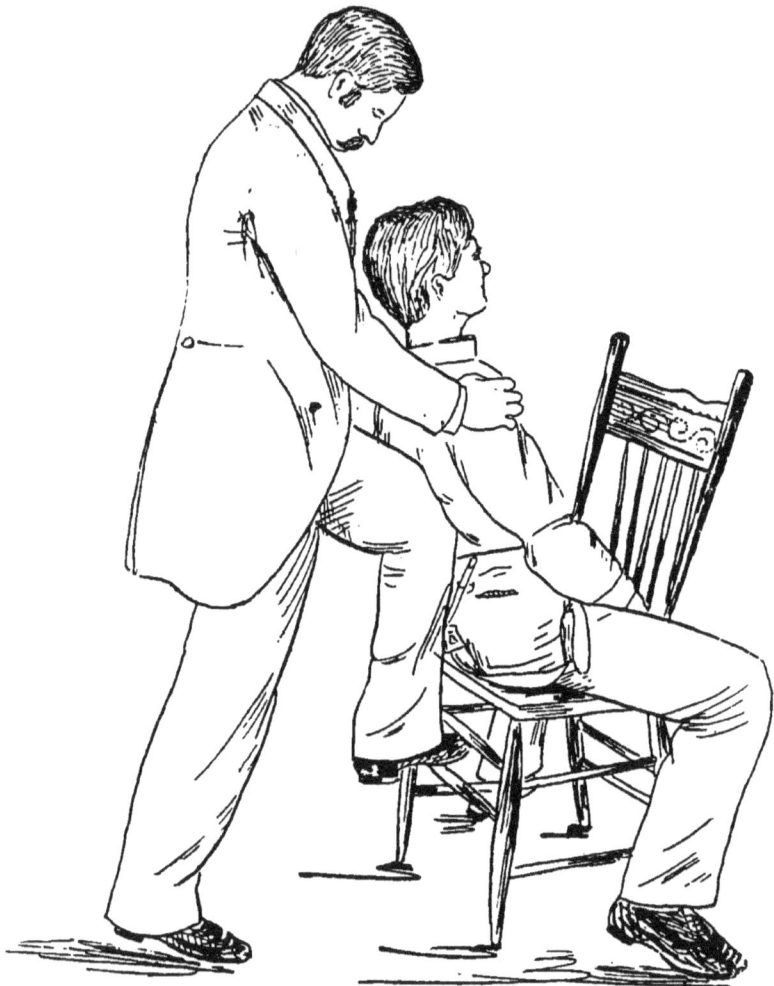

Cut 24.

both motor and sensory fibers. It supplies the organs of
voice and respiration with motor and sensory fibers, and
the pharynx, œsophagus, stomach, and heart with motor
fibers. It emerges from the cranium through the jugular
foramen, passes vertically down the neck within the sheath
of the carotid vessels, lying between the internal carotid
artery and the external jugular vein as far as the thyroid
cartilage. Thus it will be seen that it can be reached by
a strong, steady pressure on the right side of the windpipe,
as it is commonly called, in the lower part of the neck. The
right splanchnic nerve will respond to a pressure close to the
spine between the sixth and seventh ribs.

While this treatment will cure flux and cholera morbus,
and was arrived at by studying man as a machine from a
scientific standpoint, a much simpler method, producing the
same results, will be given as our infallible mode of treating these diseases.

Flux and Cholera Morbus Treatment.

1st. Place the patient on a stool, the operator standing
behind. The operator now places his knee on the left side
of the spine, just below the last rib, grasping the patient's
shoulders, and draws him gently but firmly backward as far
as possible (cut 24). Let all motions be slow, allowing the
patient time to relax the muscles. Ninety per cent of all
cases will be cured instantly by this one move. In aggravated cases, where the patient is bedfast, while lying on
the back place one hand under each side, the fingers pressing
on each side of the spine just below the last ribs, and two or
three times slowly raise the patient until only the shoulders
and pelvis touch the bed.

2d. Press lightly with the palm of the hand on the
umbilicus (and stronger as the patient becomes accustomed
to the pressure) for a minute.

3d. Hold the vaso-motor center for two or three minutes, and your patient is out of danger (see cut 5). It is very seldom that anything farther than one backward movement is necessary.

Taken suddenly with cholera morbus between St. Louis and Kansas City, the writer cured himself instantly by bending far backward over the back of the car seat. Any of our readers can do likewise.

While on this subject, we will mention the case of a lady at Miami, I. T. We received an urgent call from her husband one Tuesday morning, but, being overwhelmed with office work, it seemed impossible for us to take the time to drive twenty miles into the Indian Nation; so it was arranged that if the drugs of the local doctors failed, and she was still alive, we should drive down Friday night. We reached her bedside at midnight (Friday night), and found her just alive. We treated her once, and in a week she was walking on the streets of Miami in perfect health.

CRAMP IN THE BOWELS.

Cramp in the bowels is caused by the too rapid action of the intestines, one fold being thrown over another; this can usually be instantly cured by bending the patient far backward as in flux. In rare cases it will be found necessary to place the patient on the back and gently but firmly knead the bowels, working deep, thus freeing the parts and giving immediate relief.

Cut 8.

CRAMP IN THE STOMACH AND VOMITING.

Treatment.

1st. Bend the patient backward as in flux.

2d. Press steadily on the pit of the stomach with the palm of the hand for a moment.

3d. Place the knee between the shoulders, raising the arms high above the head (see cut 2).

4th. Permit the patient to lie on the back, and, reaching over as in cut 8, with each of the fingers close to the spine, between and a little below the scapulas (shoulder-blades), press strongly a moment, after which hold the vaso-motor center (see cut 5, page 41).

This treatment will cure the most aggravated cases, usually in a few moments. It will be observed that we are working here on the splanchnic nerves, which are in direct communication with the stomach.

CHRONIC DIARRHEA.

Symptoms.

Simple diarrhea, remaining uncured for some time, becomes chronic.

Treatment.

Our flux and cholera morbus treatment will give immediate relief, and, if continued for from three to six weeks, a complete and permanent cure. It is rarely necessary to ask the assistance of a friend in treating this disorder. Place the thumbs close to the spine, one on each side and immediately below the last rib, and while pressing hard lean as far backward as possible; move the thumbs down one inch and lean backward; repeat this until the lumbar vertebræ are reached. Treatment should be given once each day; it

will take but a moment, and has never been known to fail.
Many cases have been cured by us in a few treatments,
dating as far back as the late war.

CONSTIPATION AND TORPID LIVER.

Symptoms.

While we admit that constipation is not desirable, and
may almost invariably be avoided, yet persons thus predis-
posed are generally long-lived, unless they commit suicide
by purgative medicines, while those who are subject to
frequent attacks of diarrhea are soon debilitated. A daily
action of the bowels is no doubt desirable in most cases, but
by no means invariably so. An evacuation may take place
daily, or every second day, or even every third day, in persons who are equally healthy. There is no invariable rule
applying to all persons. Purgation produced by drugs is
an unnatural condition, and although temporary relief often
follows the use of aperients, they tend to disorganize the
parts on which their force is chiefly expended. The intestinal canal is not a smooth, hard tube, through which can be
forced whatever it contains without injury; it is part of a
living organism, and needs no force to propel its contents on
their way; nor can such force be applied with impunity. Not
only does the frequent use of purgatives overstimulate the
liver and pancreas, but also and especially the numerous
secretory glands which cover the extensive surface of the
intestinal canal, forcing them to pour out their contents in
such excessive quantities as to weaken and impair their
functions, and so produce a state of general debility. The
normal action of the stomach and intestinal canal being thus
suspended, nausea, vomiting, griping, and even fainting, are
produced; the brain and vital energies are disturbed, occa-

sioning lowness of spirits and melancholy, alternating with mental excitement and peculiar irritability of temper.

Cause.

We will now endeavor to prove to the satisfaction of our readers that, viewing man as a machine, constipation can be traced to its true cause, and cured by an application of the never-failing principles of Osteopathy. The digestive organs in constipation may be compared to an electric car with the current partially cut off; with a light load it might possibly work in a feeble, halting manner, while the slender wire transmits the power to move the heavy car. The dynamo generates that power; break the connection, and the car stops. So is the human being; the brain is the great generator, the center of all power. Stop for one instant the current on these slender nerves, and the heavy muscles of the giant are weaker than those of a tiny child. There is one peculiarity about the nerves which is liable to lead one astray, and that is the fact that a pressure on the main trunk of a nerve causes no pain at that point, but at the extremity of the nerve.

In constipation we find the intercostal and spinal muscles contracted from the fifth dorsal vertebra and fifth rib to the eighth. The sixth or seventh rib will be turned slightly, and either the muscles or rib pressing on the splanchnic nerves (which, with the pneumogastric, control the abdomen), thus depriving the intestines of half their motor power. Taking physic for constipation is like whipping a weak, half-starved horse. He will go just as long as you continue to apply the whip, but is left in a more enfeebled condition after each application of the lash. Would it not be more human and sensible to increase his feed and reduce his load, as we now propose doing with the splanchnic nerve?

By relaxing the contracted muscles we not only allow the ribs to spring back, thus releasing the nerve, but also permit the blood to pass down and supply the nerve with food, and in a comparatively short time it will be able to once more convey the current that will start the peristaltic action of the bowels, and also furnish a motor power to the sluggish liver and pancreas, enabling those organs to resume their work. As an obstruction to the nerve force of the splanchnic system not only weakens the peristaltic action of the bowels, but also the action of the liver, that great chemical laboratory; placed on the highway by which the great majority of material absorbed from the intestines reaches the blood, it is obviously in a position to act as the guardian of the blood's purity and health. It certainly in some respects performs this duty, for many poisons, when introduced into the digestive tract, are stopped by the liver, and, if their amount be not excessive, are elimated with the bile. But we have no reason to believe that this enormous mass of protoplasm is placed in this peculiar position in the circulation to preside over much more important duties than that of a mere gatekeeper. Many if not all of the absorbed materials are found to be altered during their visit to the liver. In fact, we must regard this organ as the great chemical laboratory of the blood, where many important analyses are made. It has an immense double blood-supply; it receives all the blood of the portal veins coming from the digestive tract and spleen. This supply of blood varies much in amount; after meals, it equals one-fourth of all the blood in the body. Among the many important functions of the liver are the formation of the urea and uric acid and the secretion of the bile. Its failure to supply in sufficient quantities the latter (which is mixed in the abdomen with the pancreatic juice, to assist in digesting the food) is one of the secondary causes of constipation; another is the inability of the pancreas, through lack of nerve-force, to do its part in furnishing pancreatic juice. Thus, when we turn the current on the splanchnic, we start a three-horse team, which, pulling together in perfect harmony, will safely carry our constipated friend to the highway of perfect health.

Treatment.

1st. Place the patient on the side and proceed to free all the muscles of the spine on each side as low as the twelfth dorsal vertebra. Let the arm of the patient rest on that of the operator, the patient's elbow pressing against the humerus, forming a lever with which the muscles of the scapula (shoulder-blade) can be manipulated. With the fingers between the spine and scapula, pressing hard, move the scapula and muscles under the fingers upward (see cut 3), being particular not to let the hand slip over the muscles, but to move them. After each upward motion, move the fingers down an inch, until the last dorsal vertebra is reached, taking care not to work lower than the last rib.

2d. With the patient lying on the back, grasp the right wrist with the right hand, drawing the arm slowly but with some strength high above the head, at the same time placing the left hand between the shoulder-blades on the right side of the spine, about two inches below the upper part of the shoulder-blades, pressing hard as the arm comes up; lower the arm, the elbow passing below and at the side of the table. Repeat, moving the hand down the spine one inch every time, until you have reached the tenth dorsal vertebra, which will be found one inch below the inferior angle of the scapula.

3d. Knead the bowels (cut 9), beginning on the right side and at the lower portion of the abdomen, close to the bone, and immediately over the ileo-cæcal valve. Work lightly at first, gradually using more strength, following the ascending colon upward from its commencement at the cæcum to the under surface of the liver on the right side of the gall-bladder, where it bends abruptly to the left, forming the hepatic flexure; it now becomes the transverse colon, and passes transversely across the abdomen from right to left, where it curves downward beneath the lower end of the spleen, forming the splenic flexure. The descending colon passes almost vertically downward to the upper part of the left iliac fossa, where it terminates in the sigmoid flexure. The sigmoid flexure is the narrowest part of the colon. It is situated in the left iliac fossa, commencing at the termination of the descending colon at the margin of the crest of the ilium, and ending in the rectum opposite the left sacro-iliac symphysis. Work across the abdomen, following the transverse colon and down the descending and sigmoid portions to the rectum. Next knead the small intestine, which is contained in the central lower part of the abdominal cavity, surrounded above and at the sides by the colon or large intestine.

This treatment should be given every other day, and can be administered in fifteen minutes. It will cure the most stubborn cases of constipation or torpid liver. Care should be taken to work as deep and as far under the ribs as possible. Children and young people are often cured in a single treatment, but the average time required for a cure is from two to six weeks. In very stubborn cases it is well to flush the bowels once or twice, until Nature begins to act. This treatment, if applied as directed, will be found infallible.

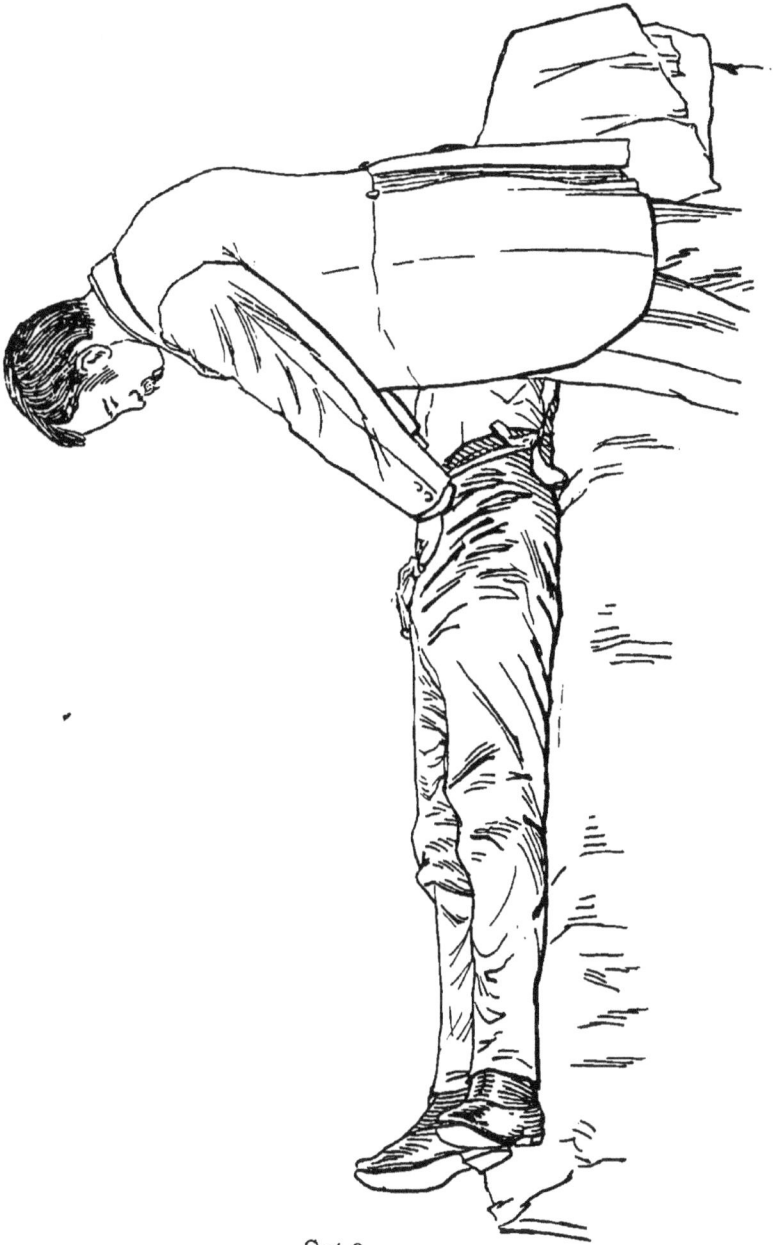

Cut 9.

DYSPEPSIA (INDIGESTION).

Symptoms.

These vary greatly, both in character and intensity, but there is commonly one or more of the following: impaired appetite, flatulence, and nausea; eructations, which often bring up bitter or acid fluids; furred tongue, often flabby, large, or indented at the sides; foul taste or breath and heartburn; pain and a sensation of weight and inconvenience or fullness after a meal; irregular action of the bowels; head-ache, diminished mental energy and alertness, and dejection of spirits; palpitation of the heart or great vessels, and various affections in other organs.

Cause.

Dyspepsia, or indigestion, is usually caused by a con-stipated condition of the bowels, which, becoming over-loaded, hinder the action of the stomach until the glands of that organ become diseased. Thus we are again confronted with the parallel of an electric car, which, having lost its current, obstructs the main track. We expect to prove to the intelligent reader that when the peristaltic action of the small intestine loses part of its nerve-power, occasioned by a pressure on the splanchnic center at or near the spine, the foundation is laid not only for constipation, liver complaint, and various stomach and kidney troubles, but by blocking the main track, one organ after another becomes diseased, and finally the stomach, bloated or filled with gas, presses so hard upon the diaphragm, compressing the left lung, that it affects even the lungs and heart. Many cases of heart trouble we have traced directly to this cause, and cured by working on these never-failing principles.

Treatment.

1st. Treat the patient as in constipation, except knead-ing the bowels, which should be omitted when constipation is not present.

2d. Stand behind the patient, and, raising the right arm high above the head, lifting strong, press hard with the left thumb on the fourth dorsal vertebra, lowering the arm with a backward motion. This pressure reaches nerves that control the pyloric valve, causing the pyloric orifice to open and permit gases and undigested food to pass into the duodenum.

Four weeks' treatment should cure any case of this complaint.

BRIGHT'S DISEASE.

Symptoms.

Dropsy of the upper as well as lower parts of the body, the hands and feet as well as face being swollen; a dry, harsh skin; quick, hard pulse; thirst, and often sickness from sym-pathy of the stomach with the kidneys. There is frequent desire to pass water, which is scanty, highly colored or smoky-looking, albuminous, and of high specific gravity.

KIDNEY DISEASE, OR INFLAMMATION OF THE KIDNEYS.

Symptoms.

Disease of the kidneys may be distinguished from colic and other similar affections by the pain being far back, and by the urine being of a deep red color, voided frequently and in small quantities. It may be distinguished from rheuma-tism by the pain not being increased by motion. This disease is attended with a sharp pain on the affected side, with much difficulty in passing urine; the bowels are costive; the skin is hot; the patient feels great uneasiness when he attempts to walk or sit upright, and lies with most ease on the affected side. Remission of the pain, discharge of high-colored mucous urine, sweating, or a flow of blood from the hemorrhoidal veins, passed in the stools, are favorable symptoms.

The Kidneys.

The kidneys, two in number, are situated in the back part of the abdomen, and are for the purpose of separating from the blood certain materials which, when dissolved in a quantity of water, also separated from the blood by the kidneys, constitutes the urine. They are placed in the loins, one on each side of the vertebral column; their upper extremity is on a level with the upper border of the twelfth dorsal vertebra, and their lower extremity on a level with the third lumbar vertebra. The right kidney is usually on a lower level than the left, probably on account of the vicinity of the liver. Each kidney is about four inches in length and two to two and one-half in breadth; they are a little over an inch in thickness. The kidney is plentifully supplied with blood by the renal artery, a large offset of the abdominal aorta. The nerves, although small, are about fifteen in number. They communicate with the spermatic plexus, a circumstance which may explain the occurrence of pain in the testicle in affections of the kidneys. It will be remembered that urea and uric acid are formed in the liver and transmitted from that organ to the kidneys. Thus it will be seen that with a diseased liver we cannot have healthy kidneys. While we have as yet been unable to cure Bright's disease, diabetes, and troubles of the bladder and kidneys caused by stricture, all other urinary troubles can be immediately cured, and those just mentioned generally benefited, by the following treatment.

Treatment for Kidney Troubles.

1st. Place the patient on the side, and, if the kidneys are inactive, begin at the sixth dorsal vertebra, with the ends of the fingers close to the spine, moving the muscles upward and outward, working deep the entire length of the spine on each side. This treatment excites the nerves to renewed action and increases the action of the blood, thus nourishing the weakened parts and giving them more strength to act.

2d. The kidneys should be kneaded gently at first, and gradually with a stronger force. It is indeed surprising how rapidly a patient will recover under this treatment.

3d. When the kidneys are too active, a steady pressure should be given from the twelfth dorsal vertebra to the last lumbar vertebra, close to the spine, in any manner most convenient to the operator.

4th. Place the patient on the face, and, while pressing hard on the sacrum immediately below the small of the back, raise the limbs from the table as high as the patient can bear without too much inconvenience, moving them gently from side to side (see cut 10). Mothers whose children have no control over the urine can cure them entirely of this annoying trouble, in one or two treatments, by pressing on the sacrum close to the last lumbar vertebra and raising the limbs as shown in cut 10.

This treatment is as reliable as mathematics. Among the many cases cured by us of this trouble we will mention our first experience in this line. A young man of Kirksville, Mo., hearing that we were investigating this subject, called at our office, and, after explaining that he had no control over his urine and had been expending all his earnings in vain hope of relief, asked us to take his case. Not having

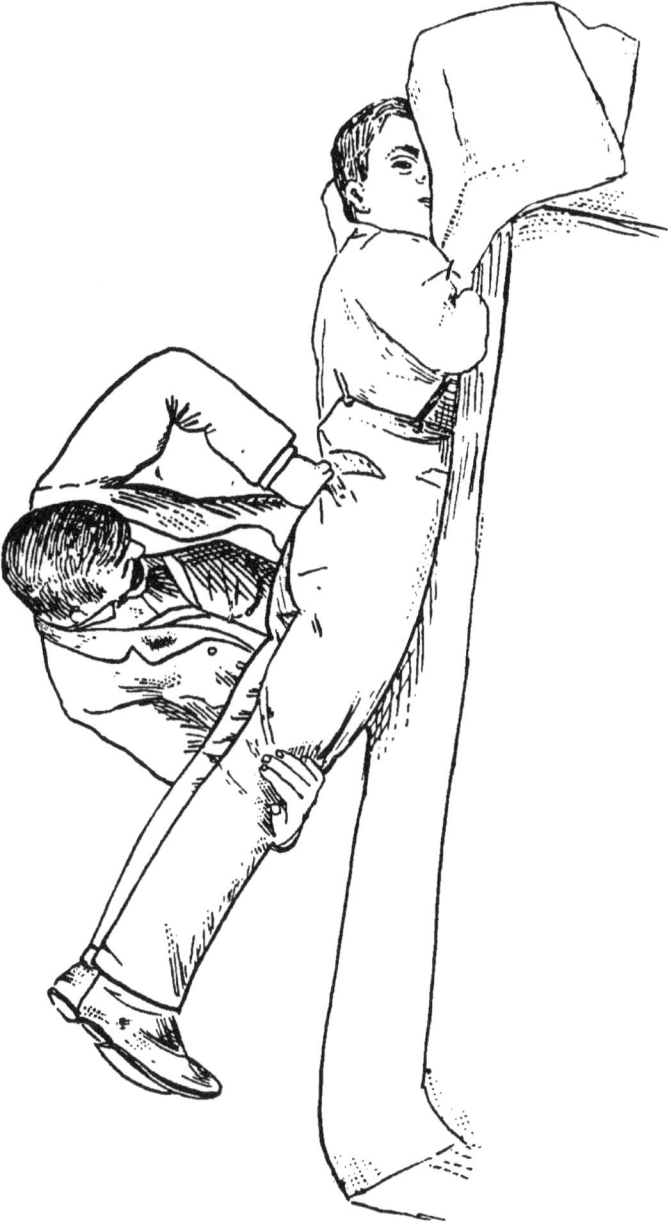

Cut 10.

a table at that time, we caused him to lie on his stomach on the floor, and, placing the right foot between his thighs and the left on the sacrum, with an ankle in either hand, we raised the limbs, sprung down the sacrum, and asked him to call again on the second day. While administering the third treatment we inquired as to results, and were not only gratified, but surprised, to learn that he had had no trouble since the first treatment. Two years later the young man was still in perfect health. We might mention also a gentleman 82 years of age, of Lewiston, Mo., troubled with this disease for over thirty years. He was entirely cured in four weeks by this method of treatment.

ENLARGED SPLEEN.

Symptoms.

This disease is characterized by a sharp or chill pain beneath the lower left ribs; with more or less tenderness on external pressure; in some instances there is very little pain, simply a feeling of weight or fullness, which is worse when the patient lies on the affected side. The attack is generally accompanied by chills and fever, and sometimes there is nausea and vomiting, cough, difficulty of breathing, and hiccough. The spleen often becomes enlarged so as to be felt beneath the lower left ribs.

The Spleen.

The spleen is situated under cover of the ribs of the left side, being separated from them by the diaphragm (the great muscle of respiration), and above by a small portion of the left lung. Its position corresponds to the ninth, tenth, and eleventh ribs. We find that enlargement of the spleen and other splenic troubles are caused by a contraction

of the muscles of these ribs and a consequent settling of one
or all of the ribs mentioned upon the spleen. Sharp pains
in this region are from the same cause, and can be instantly
relieved, and in a few weeks cured, by our treatment for
enlargement of the spleen.

Treatment.

1st. Place the patient on the right side, and loosen
all the muscles over and around the eighth, ninth, tenth,
eleventh, and twelfth ribs from the spinal column to the
median line in front, always moving the muscles upward
and not permitting the flesh to slip under the hand.

2d. Place the patient on a stool, the thumb of the right
hand on the angle of the rib at fault, and roll the arm slowly
but very strongly high above the head; lower the arm with a
backward motion; at the same instant, with the thumb of the
right hand, spring the rib forward off the spleen (see cut 11).

This treatment usually gives immediate relief, and in
from two to four weeks will effect a cure.

HEART DISEASE.

Symptoms.

The patient is seized with a sudden dreadful pain, which
centers in the heart and extends over more or less of the
anterior portion of the chest, up the shoulder, and down
the arm. There is an agonizing sense of anxiety, faintness,
and fear of instant death, palpitation, and difficulty of
breathing, so that, if walking, he is compelled to stop and
to fix on the first object that offers support, and so remains,
pale and covered with a clammy perspiration. The paroxysms may terminate in a few minutes or last for hours, and

Cut 11.

are liable to recur with increased severity till at length one proves fatal.

The Heart.

The heart is a hollow, muscular organ, of a conical form, placed between the lungs and enclosed in the cavity of the pericardium. The heart is placed obliquely in the chest, the broad attached end or base upward, backward, and to the right, and corresponds to the interval between the fifth and eighth dorsal vertebræ; the apex is directed downward, forward, and to the left, and corresponds to the space between the cartilages of the fifth and sixth ribs, three-quarters of an inch to the inner side and an inch and a half below the left nipple. The heart is placed behind the lower two-thirds of the sternum, and projects farther into the left than into the right cavity of the chest, extending from the median line about three inches in the former direction and only one and one-half inches in the latter. The heart in the adult measures five inches in length, three inches and a half in the broadest part, and two inches and a half in thickness. The average weight in the male varies from ten to twelve ounces, and in the female from eight to ten. The heart continues to increase in weight, also in length, breadth, and thickness, up to an advanced period of life. The heart of man and warm-blooded animals may be said to be made up of two muscular sacs, the pulmonary and systemic pumps, or, as they are commonly called, the right and left sides of the heart. Between these no communication exists after birth. Each of these sacs may be divided into two chambers. One, acting as an ante-chamber, receives the blood from the veins; it has very thin walls, and is called the auricle; the other, the ventricle, is the powerful muscular chamber which pumps the blood into and distends the arteries. It has been found that stimulation of the cervical portion of the spinal

cord causes quickening of the heart-beat, while a steady pressure on the same nerve-centers slows the action of the heart.
It is thus that in fever, working from this center, we slow
the heart's action, and are thereby enabled to reduce any
fever in an incredibly short time.

We will now for a moment touch upon the arteries and
veins, my object being to prove to my readers that the heart,
arteries, and veins are simply different parts of the same
machine, and that the contraction of a muscle, throwing a
pressure on an artery or vein, will affect the heart, on the
same principle that a force-pump attached to a rubber hose
would be affected should you stand on the hose. The channels which carry the blood through the body form a closed
system of elastic tubes, which may be divided into three
varieties: arteries, capillaries, and veins. The arteries are
those vessels that carry the blood from the heart to the
capillaries. The great trunk of the aorta springs from the
left ventricle and gives off a series of branches, which in
turn subdivide more and more freely in proportion to their
distance from the heart. The aorta is divided into the arch,
ascending and descending portions. The descending aorta
is divided into two portions, the thoracic and abdominal,
in correspondence with the two great cavities of the trunk
in which it is situated. The thoracic aorta commences at
the lower border of the fourth dorsal vertebra on the left
side, and terminates at the aortic opening in the diaphragm,
in front of the last dorsal vertebra. The abdominal aorta
commences at the aortic opening in the diaphragm, in front
of the body of the last dorsal vertebra, and, descending a
little to the left of the vertebral column, terminates opposite
the body of the fourth lumbar vertebra, where it divides into
the right and left common iliac arteries. The common
iliac arteries are about two inches in length, and divide
opposite the intervertebral substance of the last lumbar

vertebra and sacrum, into the internal and external iliac arteries, the latter supplying the lower extremities. The external iliac artery passes obliquely downward and outward along the inner border of the psoas muscle from the bifurcation of the common iliac arteries to Poupart's ligament, where it enters the thigh and becomes the femoral artery. The femoral artery commences immediately behind Poupart's ligament, midway between the anterior and superior spine of the ilium and the symphysis pubis, and, passing down the front and inner part of the thigh, terminates at the opening of the adductor magnus muscle at the junction of the middle with the lower third of the thigh, where it becomes the popliteal artery. The popliteal artery commences at the termination of the femoral at the opening in the adductor magnus, and, passing obliquely downward and outward behind the knee-joint to the lower border of the popliteus muscle, divides into the anterior and posterior tibial arteries. The anterior tibial artery passes forward between the two heads of the tibialus posticus to the deep part of the front of the leg; then descends on the anterior surface of the interosseous membrane, gradually approaching the tibia, and at the lower part of the leg lies on the bone, and then on the anterior ligament of the ankle to the bend of the ankle-joint, where it lies more superficially and becomes the dorsalis pedis. This artery passes forward from the bend of the ankle along the tibial side of the foot, and terminates in two small branches, the dorsalis hallucis and communicating. The posterior tibial artery is of large size, and passes obliquely downward from the lower border of the popliteus muscle along the tibial side of the leg to the fossa between the ankle and the heel, where it divides into the internal and external plantar arteries.

Now, having traced this river of blood, which throws

branches to each organ and muscle in its course from the heart to its termination in the lower extremities, passing as it does through, over, under, and between the numerous muscles on its journey, it will not be hard for the intelligent reader to believe that an obstruction to its free flow, caused by contracted muscles, would affect the heart. Cramping of the muscles is so very common, often leaving the muscles in ridges, that the most skeptical will not dispute the fact that muscles will contract and remain in that condition. The heavy muscles of the thigh and those below and about the knee, from their peculiar relation to the artery and their great strength, are usually at fault, and by a simple twist of the leg, throwing these muscles on a strain, and thereby freeing the femoral artery, we have cured cases of heart disease that had baffled the best physicians of modern times.

Treatment.

First ascertain if the patient has cold extremities; such being the case, the trouble is necessarily along the line of the artery, and the heart trouble corresponds to the illustration of pump and rubber hose when the hose is obstructed.

1st. Place the patient on the back, and, grasping the ankle in the right hand, flex the leg against the chest, slowly but strongly; while pressing it hard against the chest, rotate if from right to left as in cut 6. Bring it to its full length, turning the knee in and foot out with a light jerk (cut 7). This should be repeated with each limb two or three times at each treatment.

2d. Flex the knees, place the feet together, and, with one hand on each knee, spread the knees as far apart as possible, thus stretching the adductor muscles (see cut 12).

3d. Placing one hand on each side of the thigh, close to the body, move all the flesh, very deep, from right to left, the entire length of the limb (cut 13).

Cut 12.

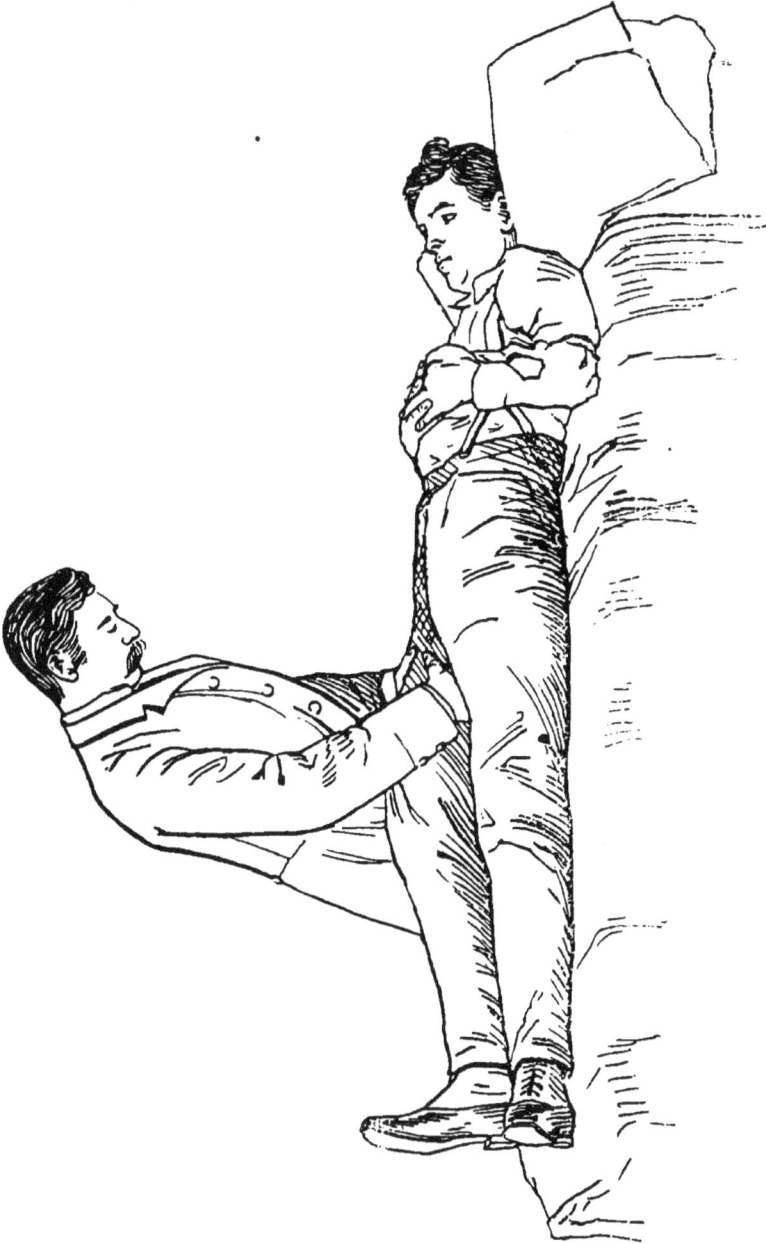

Cut 13.

4th. Grasp the foot, and, while rotating the limb, pull slowly but strongly.

We guarantee that any case of cold limbs accompanied by heart trouble, or in which the organic difficulty is not yet perceptible, can be cured in from two to four weeks by this treatment, which should not take over ten minutes each time, and should be given every other day.

"ENLARGEMENT OF THE HEART," ETC.

Heart disease is often caused by constipation and a diseased condition of the alimentary canal. Those cases can be readily distinguished, as the patient will find great difficulty in breathing when in a recumbent position, and upon resuming an upright position will feel immediate relief, thus proving that the organs are distended to such an extent that there is at all times a pressure on the diaphragm left lung, and heart, and that their own weight when in an upright position will partially free the last named organs. It will be readily understood that out constipation or indigestion treatment, or both, as the case may seem to require, will cure this form of heart disease, as it has done in hundreds of cases. Other cases, and they are very numerous, are caused by a contraction of the muscles, depressing the ribs immediately over the heart, thus interfering with its action. We are led to believe that there are very few cases of actual "enlargement of the heart," but that the so-called "enlargement of the heart" is really a compression of the cavity in which the heart is contained.

Treatment for "Enlargement of the Heart."

1st. Free all the muscles attached to the ribs immediately over the heart, from the spine to the median line, on each side, always moving the flesh upward, using the arm as a lever in treating the muscles of the spine (cut 3).

2d. Place the patient on the back, one operator grasping each wrist; placing the disengaged hand between the patient's shoulders, the fingers pressing hard upon the angle of the rib between the spine and scapula, draw the arms slowly, but with some strength, high above the head; move the hands down one inch, and repeat until you have reached the lower angle of the scapula.

This will usually give instant relief, and never fails to effect a cure in from two to four weeks' treatment. A treatment should be given every other day.

Of the many cases cured by us in this manner, it might be well to mention an old gentleman of Galena, Kansas. As a drowning man will grasp at a straw, when he was dying and nearly all hope had fled, we were called in, and, in the presence of two medical doctors and the members of his family, we raised his ribs, thus permitting the heart to act. In a few minutes we had our patient out of danger. We will also add that he never after had any difficulty with his heart.

, FEEBLE ACTION OF THE HEART.

This trouble is caused by an almost imperceptible contraction of all the muscles, thus interfering with the entire circulation. An aching, tired sensation, so often felt, is caused by the contracting muscles, as is readily proven beyond the shadow of a doubt by the fact that after a general treatment, stretching and moving all the muscles, permitting the sluggish blood to move more rapidly through the arteries and veins, the heart's action is increased and the tired, aching, worn sensation has entirely disappeared.

General Treatment.

1st. Place the patient on the side, using the arm as a lever (cut 3), with the fingers pressing rather hard close to the spine; beginning at the first dorsal vertebra, free all the muscles the entire length of the spinal column. Have the patient turn over, and treat the other side in a similar manner.

2d. Flex the limbs against the chest, the patient lying on the back; rotate the leg from right to left two or three times, extending with a light jerk (cuts 6 and 7).

3d. Grasp the thighs firmly with one hand on each side, the fingers meeting; beginning close to the body, move the flesh, to the bone if possible, from right to left, the entire length of the limb (cut 13).

4th. Place the patient on a stool, the operator standing at his back, and, placing the knee between patient's shoulders, grasp the wrists and raise the arms strongly high above the head (see cut 2); at this time the patient will inhale, filling the lungs with air; lower the arms with a backward motion.

5th. Place one hand under the chin, the other under back of the head, and pull gently, rotating the head from side to side (cut 4).

6th. Placing one hand on each side of the neck, the fingers almost meeting at the back of the neck, close to the head and the breadth of four fingers down the neck, press gently on the vaso-motor center (see cut 5) for a few moments, to quiet the nerves.

This treatment will require not over fifteen minutes, and should be given every other day. We guarantee it to give immediate relief, and, if continued for from two to four weeks, a permanent cure will be effected.

Of the numerous cases cured by our general treatment, we will mention that of an old gentleman brought to us in

Galena, Kansas. He could not climb the steps to our office. We treated him on the counter in a grocery store near by. His pulse, which was hardly perceptible, was down to 38. When he came for his second treatment, two days later, his pulse was strong and had increased to 58. He could hear better, and recognized people on the street for the first time in months

DROPSY.

Symptoms.

Dropsy is watery accumulation in the areolar tissue more or less generally throughout the body. It is of two distinct varieties, for, besides its occurrence in the meshes of the loose tissue beneath the skin, it may take place as a local dropsy in any of the natural cavities or sacs of the body, and is named according to the parts involved.

Partial dropsy is always due to excessive venous repletion, and this overdistention of the small veins is the result of some mechanical impediment to the venous circulation. Dropsy due to obstructed portal circulation may be recognized by the following clinical characters: It begins in the abdomen; difficult breathing follows, but does not precede the ascites. There is a tendency to vomiting, diarrhea, and piles; further, the spleen becomes enlarged and there are varicose veins on the right side of the abdomen.

Dropsy at first partial, but afterwards becoming general, commences in the feet and extends upward, and this is also due to excessive venous repletion from obstructed venous circulation.

Dropsical swellings are soft, inelastic, diffused, and leave, for some time, the indentation made by the pressure of a finger. In chronic cases and when the swelling is very great the skin becomes smooth, glassy, and of a dull red or purple color, and where the skin is less elastic it becomes livid or blackish and troublesome, even gangrenous, or sloughs may form.

In treating dropsy, of whatever organ, it is necessary to use such remedies as will act on the kidneys and skin and excite them to increased activity; the result of this activity is to diminish the fluids which have collected in one or another part of the body and remain there unabsorbed, and cause them to be taken up by the kidneys or thrown off by the skin, and thus carried out of the system through the natural outlets. Any remedy that accomplishes this object effectively cures dropsy occuring in any part of the body.

Cause of Dropsy.

That the reader may gain a more correct understanding of our method of treating dropsy, viewing the human body as a machine, we will once more refer to the anatomy. Having followed this river of blood from the heart to its termination, we must now trace it back to the heart and endeavor to locate along its channel the cause of dropsy and consequent heart trouble. The frequently branching arteries finally terminate in the capillaries, in which distinct branches can no longer be recognized, but their channels are interwoven into a network, the meshes of which are made up of vessels all having the same caliber. They communicate with the capillary network of the neighboring arteries, so that any given capillary area appears to be one continuous net of tubules connected here and there with a similar network from distant arterioles, and thus any given capillary area may be fed with blood from several different sources.

Veins.

The veins arise from the capillary network, commencing as radicles, which correspond to the ultimate distribution of the arterioles, but they soon form wider and more numerous channels. They rapidly congregate together, making comparatively large vessels, which frequently intercommunicate and form coarse and irregular flexures. Thus it will be seen that we have two rivers, one distributing, the other gathering up and returning the blood to the heart. While a pressure on an artery, cutting off the supply to the extremities, causes them to be cold, at the same time affecting the heart, a pressure on a vein, stopping the return current, will necessitate an engorgement of the blood in the capillaries; the heart, working against heavy odds in trying to force the blood past the contracted muscles, will certainly be affected, while the stagnant blood, unable to escape, will cause either inflammatory rheumatism, dropsy, or erysipelas.

Treatment.

1st. Place the patient on the side and move all the muscles of the spine very deep from the tenth dorsal to the last sacral vertebra. This will excite the nerves which control the kidneys to renewed action, thereby enabling them to separate the immense amount of water about to be poured into them from the blood.

2d. Give general treatment (page 32), being very careful to stretch all the muscles near the affected parts.

In a very short time, usually from two to six days, the kidneys will begin to act very freely, throwing off the decomposed and watery particles of blood, while in from two to six weeks the patient will be entirely well.

Of numerous cases cured by us of dropsy we will mention that of a lady of Joplin, Missouri, whose case had not only been treated by the best physicians of her own city, but those of Kansas City and St. Louis. She came to us in a hopeless condition; her abdomen, limbs, and feet were swollen to more than twice their normal size. After the second general treatment, she began to improve rapidly, and in ten days her ankle could be spanned by the thumb and fingers; in one month the dropsy had entirely disappeared. She gained strength rapidly, and in a short time had entirely regained her health. •

Another remarkable case was that of an old gentleman of Baxter Springs, Kansas, who had been for five years gradually losing the use of his lower limbs, and during the last year dropsy had made its appearance. After the second general treatment, the dropsy had almost entirely disappeared; his limbs regained their long-lost strength, and he would leap about the office like a boy in an ecstasy of delight, kicking higher than the doctor's head and springing from the floor to our operating-table with apparent ease.

GENERAL DEBILITY.

While we cannot roll back the vail of years, we propose to prove to our friends in advanced life that we can at least make them feel quite young again. In old age the muscles, arteries, and, in fact, all the organs, are prone to ossify. The muscles become contracted and stiff, thus interfering with the free flow of blood, and limy deposits form around the joints. It is but reasonable and natural that our general treatment (page 32) which stretches and frees all the muscles, ligaments, and joints, causing the blood to run faster and the heart to beat stronger, would be especially applicable. It has been tested and proven times without number, and we feel that we can safely assert without fear of contradiction that our general treatment with the aged and infirm never has and never will fail to give gratifying results.

We will mention in this connection the case of a gentleman eighty-one years of age, of Miami, Indian Territory, in whom the machinery of life had nearly run down. His sons, hearing of some of our rather remarkable cures, brought the old man in, much against his will, for treatment. He was carried into our office and laid upon the table. After an examination, we pronounced the case hopeless, as we did not think there was enough vitality left to respond to the treatment and once more resume control of the machinery of life. However, we administered a general treatment, and were as much surprised as his sons to see the old gentleman get up and walk down stairs unassisted. In three weeks he was restored to health and threw away the cane he had carried for thirty years. Being a man of undoubted veracity and well known throughout the Indian Territory and southern Kansas, his seemingly miraculous restoration to health through this method gave us quite an enviable reputation through that section of the country. Sometimes, when weary and annoyed by many questions put to him regarding our method, he would tell the people that he was no walking advertisement.

That the young as well as the old can be benefited by this general treatment has been proved in numerous instances, after all other known methods have failed. One case we will mention is that of a child eighteen months old. Her cold, emaciated limbs, and the eruptions on her face and neck and in the ears, told too plainly to be misunderstood the story of contracted muscles and of stagnant blood. Although the little sufferer was so low that her case seemed almost hopeless, she was cured in four treatments, given every second day, and is now a healthy child.

FEVER AND AGUE.

Symptoms.

This disease may be divided into three stages: the cold stage, the hot stage, the sweating stage.

Cold Stage.—An intermitting fever generally begins with pain in the head and loins, weariness of the limbs, coldness of the extremities, stretching, yawning, with sometimes great sickness and vomiting, to which succeed shivering and violent shaking; respiration is short, frequent, and anxious.

Hot Stage.—After a longer or shorter continuance of shivering, the heat of the body gradually returns—irregularly at first, and by transient flushes, soon succeeded by a steady dry and burning heat, considerably augmenting above the natural standard; the skin, which before was pale and constricted, becomes swollen, tense, and red; pain is felt in the head and various parts of the body; the pulse is quick, strong, and hard; the tongue white, the thirst great, and the urine high-colored.

Sweating Stage.—A moisture is observed, the heat falls, the pulse diminishes and becomes full and free, and all the functions are restored to their natural order.

CONGESTIVE CHILLS.

Symptoms.

These are only an aggravated form of chills and fever, and are sometimes called "sinking chills."

We guarantee our general treatment to cure chills and fever and congestive chills in from two to six days, and to give immediate relief in all cases, even after all other known methods have failed. The almost imperceptible contraction

of muscles not only checks the warm blood, thus producing
the chill, but also causes the bones to ache. The patient
will unconsciously stretch, thus gaining momentary relief.
When the muscles finally relax, freeing the pent-up blood,
it rushes to the head and through the arteries too rapidly;
thus the accompanying fever is produced.

Treatment.

Give general treatment (page 32), if possible, just before
the chill; hold the vaso-motor center (cut 5), which causes
the arteries to contract, thus slowing the heart's action, only
when fever is perceptible. In ordinary cases there will be
but one light chill after the first treatment.

BRAIN FEVER.

Symptoms.

The symptoms which usually precede brain fever are
pain in the head, redness of the eyes, a violent flushing of
the face, disturbed sleep or a total want of it, great dryness
of the skin, costiveness, retention of the urine, a small dropping of blood from the nose, singing in the ears, and extreme
sensibility of the nervous system. The pulse is often weak,
irregular, and trembling, but sometimes is hard and contracted; a remarkable quickness of hearing is a common
symptom of this disease, as is also a great throbbing of the
arteries in the neck and temples; a constant trembling, suppression of the urine, a total want of sleep, and a grinding of
the teeth, which may be considered as a kind of convulsion.

The Brain.

Referring again to the anatomy, we find that the brain,
the great dynamo which generates the forces that control
the system, is contained in the cavity of the cranium, and,

to perform the varied tasks imposed upon it, must be bounti-
fully supplied with arterial blood, which must circulate freely
and return quickly to the heart through an unobstructed
channel. The blood leaves the arch of the aorta through
the innominate artery, and ascends obliquely to the upper‑
border of the right sterno-clavicular articulation, where it
divides into the right common carotid and right subclavian
arteries, the latter supplying the right arm, while the common carotid passes obliquely upward from behind the sterno-
clavicular articulation to a level with the upper border of
the thyroid cartilage, opposite the third cervical vertebra,
where it divides into the external and internal carotid, whose
branches, together with the vertebral artery, supply the
brain. As these arteries and the corresponding veins must
pass through a network of muscles to reach their destination,
the great mystery is that we are ever free from headache
caused by an obstruction to their free flow.

Causes of Brain Fever.

Brain fever, usually fatal when treated by the old
methods, can be traced directly to a contraction of the muscles of the neck, obstructing the returning blood. With
the heart still pumping the blood into the brain and the
escape cut off, do you wonder at brain fever, or can you
doubt for an instant that to remove the obstruction, allowing the pent-up venous blood to escape down its proper
channel, would cure the disease?

We have yet to find a case of brain fever that, if taken
in any reasonable time, can not be instantly relieved, and
in a comparatively short time cured, by our brain fever
treatment. We trust that not only every person who reads
these pages, but the medical fraternity in particular, will
try this method, as, if it is generally adopted, it will save
hundreds of lives annually.

Treatment.

1st. Place one hand under the chin, the other under the back of the head, and pull gently, rotating the head as far as possible from side to side, the object being to stretch all the muscles of the neck (see cut 4).

2d. Pull gently on the head (being very careful not to rotate it) until sufficient strength is used to move the body.

3d. With the fingers move all the flesh and muscles of the neck and throat, working gently but deep.

4th. Raise the arm high above the head (see cut 14) with one hand, with the fingers of the other pressing hard between the spine and scapula (shoulder-blade), beginning at the upper border of the scapula. Lower the arm with a backward motion, and repeat, moving the fingers down one inch each time until the lower angle of the scapula is reached. Treat the other side in a similar manner.

5th. Hold the vaso-motor center, one hand on each side of the neck, the fingers almost meeting close to the head (cut 5), and in five minutes your patient will be asleep and out of danger. In critical cases this treatment may be repeated as often as circumstances seem to require. Once in six hours is usually all that is necessary. The vaso-motor center may be held at any time, and always gives relief.

Of the many cases treated by us, we will mention that of a gentleman of Baxter Springs, Kansas, who was delirious when we reached his bedside, and had been given up by the family physician. After a treatment that lasted not over ten minutes, his wife, bending over the couch, said to him: "Ben, how do you feel?" He replied: "Better, you bet!" turned over, and went to sleep for the first time in days. He improved rapidly, and in two weeks had entirely recovered.

We are perfectly satisfied that brain fever, if treated in time by these never-failing principles, is no more to be dreaded than a bad cold.

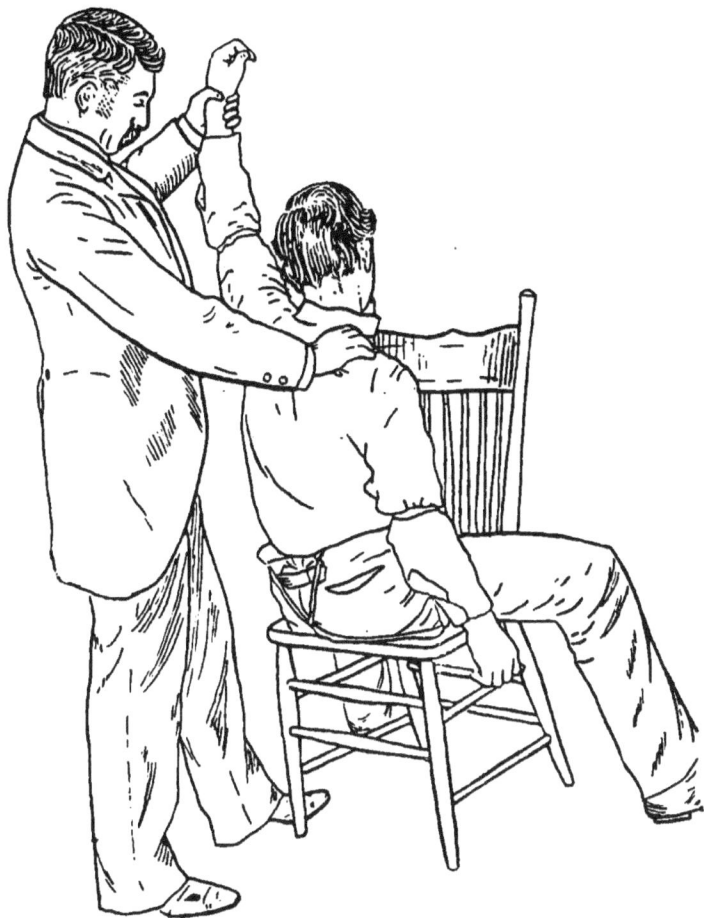

Cut 14.

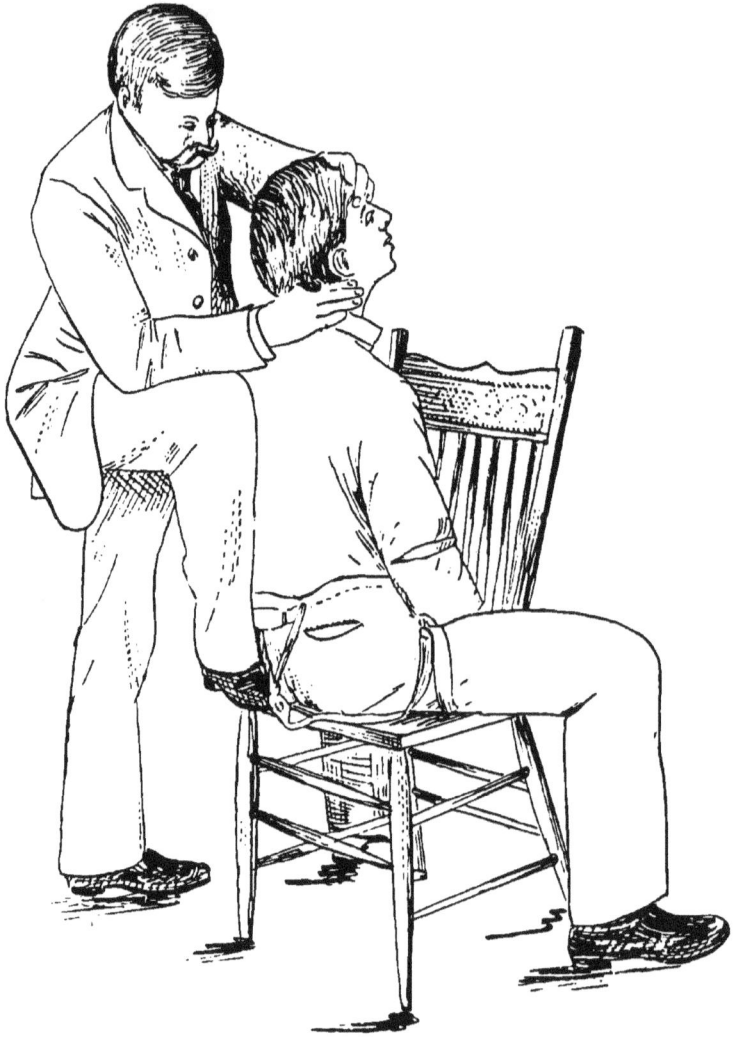

Cut 15.

HEADACHE.

Acting on these great principles, headache, not caused by fevers, the stomach, or the uterus, can be almost instantly cured by stretching the neck and a pressure on the nerves at the base of the occipital bone.

Treatment.

1st. Place the right hand on the back of the patient's neck, the thumb on one side and the fingers on the other, close to the head; place the left hand on the forehead, tipping the head backward, gently, lifting quite strongly with the right arm, while rotating the head gently from side to side (see cut 15).

2d. Standing in front of the patient and tilting the head backward, gently hold the vaso-motor center (cut 16).

3d. Place one hand on the forehead, the other upon the back of the head, and press for a moment, hard.

4th. Standing behind the patient, with one hand on each side of the forehead, make five or six quick strokes.

5th. Place one finger on each temple, and, while pressing, gently move the fingers from right to left with a circular motion.

6th. With the right hand raise patient's right arm high above the head, with the thumb of the left hand between the spine and the scapula, beginning at its upper angle, moving the muscles upward with a strong pressure at the same instant (see cut 14).

This treatment will not occupy over ten minutes. First, second, and third will cure any ordinary headache, and the entire treatment will cure any case of headache, no matter how severe, if not caused by fevers, the stomach, or the uterus, in from five to ten minutes.

Of the scores of cases cured by us, we will mention the

case of a lady of Scammon, Kansas, who would have an attack of nervous headache once each week, lasting from two to four days; she suffered intense pain, the muscles contracting until they drew the head backward upon the neck. At these times electricity, hypodermic injections, and even chloroform, had failed to give relief. Being anxious to put our method to the severest test, we undertook her case, and were even more successful than we had dared to hope. In ten minutes she was sleeping quietly, apparently free from pain. In the morning a second light treatment was given, which gave her complete relief, and the continuation of the treatments for one month effected a permanent cure.

NERVOUS HEADACHE.

Treatment.

In addition to our headache treatment, place the patient on the face, and with the thumbs, beginning at the neck, press gently the entire length of the dorsal vertebræ, after which, if the patient has retired, hold the vaso-motor center a moment (cut 5), and in a few minutes your patient will be asleep.

SICK HEADACHE.

Sick headache being caused by a reflex action of the pneumogastric and splanchnic nerves, the stomach is at fault; we must first reach that organ through the splanchnic nerves.

Treatment.

1st. Place the patient on the back, and, reaching over as in cut 8, with the fingers pressing hard on each side of the spine, beginning between the lower angle of the scapula and ending as low as the last dorsal vertebra, lift the patient

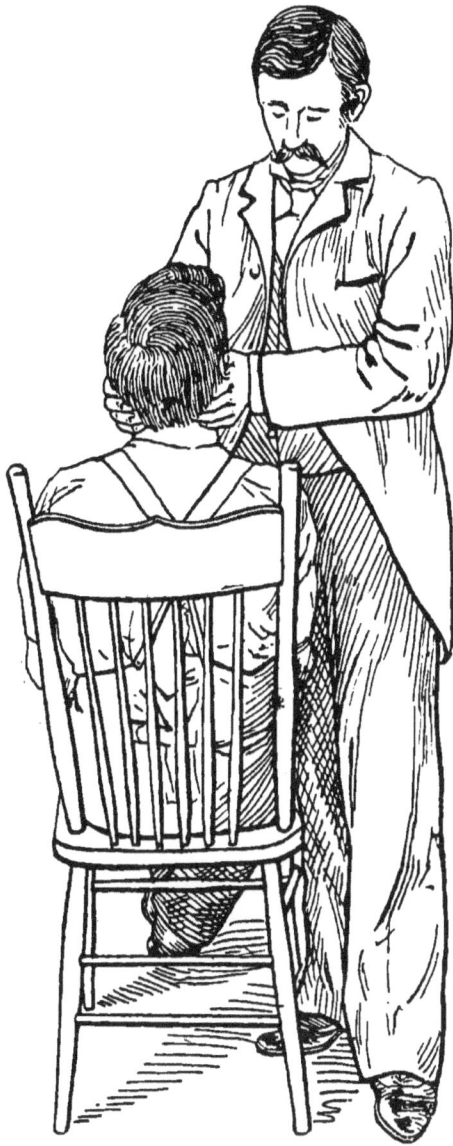

Cut 16.

gently with your fingers, then, moving down the breadth of the hands, repeat the application, thus desensitizing the splanchnic nerves.

2d. Press gently at first, then gradually harder, over the pit of the stomach.

3d. Give our regular headache treatment. It will take from ten to fifteen minutes to give this treatment, at which time the patient will be improving, although it may be some time before the pain entirely abates.

A continuation of this treatment every second day will cure the most aggravated cases of chronic sick headache.

Pain in the top of the head in women is always caused by female troubles, and will be discussed at length under that head.

THE DIAPHRAGM.

The diaphragm is the principal muscle of inspiration, placed obliquely at the junction of the upper with the middle third of the trunk, and separating the thorax from the abdomen, forming the floor of the former cavity and the roof of the latter. When in a condition of rest, the muscle presents a domed surface with the concave toward the abdomen; when the fibers contract, they become less arched, or nearly straight, and in consequence the level of the chest wall is lowered, the vertical diameter of the chest being proportionally increased, thus permitting the lungs to fill with air; when, at the end of the inspiration, the diaphragm relaxes, the thoracic walls return to their natural position in consequence of their elastic reaction and of the elasticity and weight of the displaced viscera. In all expulsive acts the diaphragm is called into action; thus before sneezing, coughing, laughing, crying, or vomiting a deep inspiration takes place. The phrenic nerve, which may be reached by a pressure in front of the third, fourth, and fifth cervical vertebræ, controls the action of the diaphragm. Thus viewing that great muscle of inspiration as a machine, whose action is controlled by the brain through the phrenic nerve, we are enabled to instantly cure hiccough.

HICCOUGH.

Hiccough is a too rapid and spasmodic action of the diaphragm, arising from any cause that irritates its nervous fibers.

Treatment.

1st. Stand behind the patient and with the fingers of each hand push the muscles at the side of the neck forward and press gently on the front of the tranverse processes of the third, fourth, and fifth cervical vertebræ.

2d. Place the knee between the patient's shoulders and raise the arms high above the head, lifting strongly (cut 2).

This will instantly cure any case of hiccough not caused by approaching death.

FEVERS.

Fevers in any of their various forms can be greatly benefited and usually cured by an application of some or all of the following principles, as the case seems to require:

1st. Should the patient be constipated, flush the bowels and give constipation treatment.

2d. Should there be diarrhea, the bowels should be checked as in flux.

3d. Often a general treatment gives immediate relief.

4th. When there is a pain in the head, the headache treatment should be always given.

5th. Always in fever hold the vaso-motor center (cut 5) a few moments, as this never fails to reduce the fever and give immediate relief.

While a general treatment should not be given oftener than once each day, the vaso-motor center may be held a few moments at any time. This part of the treatment should never be omitted, even if a physician is called, as it will in no way interfere with his medicine and gives more relief than all his drugs.

HAY FEVER (SUMMER CATARRH).

Symptoms.

The symptoms are those of an ordinary catarrh, to which those of asthma are superadded. There is itching of the forehead, eyes, nose, and ears, much general irritability and lassitude, sneezing, profuse discharge from the nose, tightness of chest, difficult breathing and cough, pricking sensations in the throat, general depression, etc.

Treatment.

1st. General treatment of the neck.

2d. Asthma treatment.

Treatment should be given every other day for a few days, when the hay fever will entirely disappear.

ACUTE RHEUMATISM.

Symptoms.

Acute rheumatism is usually ushered in with fever and inflammation about one or more of the larger joints, the shoulder, elbow, knee, or ankle usually being first affected. Exposed joints appear to be more prone to attacks than those that are covered, the larger more frequently than the smaller, and the small joints of the hand more frequently than those of the feet. The affected joints are swollen, surrounded by a rose-colored blush, and acutely painful; the pain has many degrees of intensity, generally abates somewhat in the day, but is aggravated at night, and in all cases is increased by pressure, so that the touch of the nurse or weight of the bedclothes can scarcely be borne.

CHRONIC RHEUMATISM.

This sometimes follows the acute form, at other times coming on quite independently of any previous attack. In time the affected limbs lose their power of motion and lameness results, the hip- and knee-joints being most often affected. Sometimes there is emaciation of the muscles, sometimes permanent contraction of a limb or bony stiffness of the joints. This form of the disease is the result of the uncured acute form; it may be limited to one part of the body or extend to several, and may be fixed or shifting.

Cause of Rheumatism.

Rheumatism in its varied forms yields readily to this method of treatment, and I think I am justified in saying that there never was and never will be a case of rheumatism that could not be benefited by this treatment, and permanently and quickly cured if taken in any reasonable time.

Inflammatory rheumatism of the limbs is caused by a contraction of the muscles of the thigh, obstructing either the femoral, iliac, or long saphenous veins. As the waters of a river, if obstructed between high, strong banks, may not cause any particular trouble at that point, but will back up, flooding the lower country, so this river of blood, while causing no great inconvenience at the obstruction, backs up, distending the smaller veins and capillaries. The heart, still pumping, finally feels the pressure, and we have rheumatism of the heart, while the stagnant blood soon becomes feverish, and we have inflammatory rheumatism, first below and finally above the obstruction. Laying aside all prejudice and skepticism, and looking at the matter from an unbiased and common-sense standpoint, would it not be as sensible to throw medicine into the river to remove the cause of a flood as to put it into the stomach to free the obstructed vein?

If we may be permitted to once more refer to the anatomy, we will turn to the stomach and endeavor to follow a dose of medicine on its journey from the stomach to its destination at the femoral or saphenous vein, first asserting, however, that the same quantity of the same medicine never has the same effect twice on the same individual, from the fact that the stomach is never found twice in exactly the same condition, containing as it does different foods in different stages of digestion. Thus your doctor must begin by guessing what to give and how much to give, and continue guessing until you accidentally stretch the contracted muscle and commence to recover, and then he guesses that his muscle did the work. There will be no guesswork, however, in tracing the medicine from the stomach and asserting what portion (if any) reaches the affected part. Passing from the mouth through the œsophagus, it first reaches the cardiac end of the stomach. While in the mouth the gastric juice commences to flow, and is greatly increased by the time the drug gets to the stomach. Being kept in motion in a large quantity of liquid, in from fifteen to thirty minutes it reaches the pyloric orifice of the stomach and is emptied into the duodenum, where it is mixed with the pancreatic juice from the pancreas and the bile from the liver. As these juices, together with the gastric juice of the stomach, are

capable of changing the entire character of almost any substance on which they are allowed to act, it is not only
possible, but probable, that they also change the character
of the drug to a greater or less extent, thus adding to the
system of guessing indulged in by the medical practitioner.
The duodenum and upper portion of the small intestine are
lined with a velvety substance, termed *villi*, which, immediately upon the entrance of any substance into the intestine,
passes all particles of richness through the walls of the
intestine into the thoracic duct, permitting all refuse matter
to pass on to the rectum. It is possible that here a goodly
portion of the drug is refused by the villi, and is passed with
the fæces; a portion, however, has reached the thoracic duct.
The thoracic duct conveys the great mass of lymph and
chyle into the blood. It varies in length from fifteen to
eighteen inches in the adult, and extends from the second
lumbar vertebra to the root of the neck; it commences in
the abdomen by a triangular dilatation, the receptaculum
chyli, which is situated upon the front of the body of the
second lumbar vertebra, to the right side and behind the
aorta, it ascends into the thorax through the aortic opening
in the diaphragm, opposite the fourth dorsal vertebra; it
inclines toward the left side and ascends behind the arch of
the aorta to the left side of the œsophagus, and behind the
first portion of the left subclavian artery to the upper orifice
of the thorax, opposite the seventh cervical vertebra; it now
curves outward and then downward over the subclavian
artery, and terminates in the left subclavian vein.

We will now follow what remains of our drug through
the thoracic duct and into the subclavian vein in the neck,
opposite the seventh cervical vertebra. It next passes into
the innominate artery, which empties into the superior vena
cava, through which it reaches the heart, and is immediately

pumped through the pulmonary artery into the lungs, from which it is conveyed through the pulmonary veins back to the heart, from where it is distributed equally to all parts of the system. How much (if any) ever reaches the con-tracted muscle it will indeed be difficult to determine.

In discussing this subject we are reminded of the story of an Irishman with rheumatism; his physician wrote him a prescription and instructed him to rub it on. The Irish-man, in his ignorance, rubbed his leg with the paper, and he was immediately relieved; thus demonstrating that there is more virtue in the rubbing advised in the liniments than in the liniments themselves.

Now, believing that our readers will understand our reasons for taking the position that man is a machine and should be treated accordingly, especially in rheumatic troubles, we will return to the treatment of this disease when located in the lower extremities.

INFLAMMATORY RHEUMATISM.

Treatment.

1st. Place the patient on the back, and, grasping the ankle firmly with the right hand (should the right limb be affected), place the left on the knee and flex the limb slowly and gently as far as possible without too much suffering. Rotating it gently from right to left, extend the leg, and it will be found that it can be returned to its former position with apparent ease; bend it now another inch and straighten.

2d. Place one hand on each side of the thigh, close to the body, and with a firm pressure move all the muscles from right to left and from left to right (see cut 13) the entire length of the limb, very gently at first, but stronger as the patient becomes used to the treatment.

3d. Grasping the foot, pull slowly, at the same time rotating the limb, using as much strength as the patient can stand.

4th. Place the patient on the side, and, beginning at the first lumbar vertebra, with the fingers close to the spine, move the muscles upward and outward down as low as the lower border of the sacrum.

This treatment should be given every other day, and, if care is taken, it need not be very painful, and will certainly cure the most acute case of inflammatory rheumatism in from two to six weeks.

Of the numerous cases cured by us, we will mention that of a gentleman at Webb City, Missouri, who had been given up by the medical doctors and in whose case the rheumatism not only extended the entire length of the spinal column and right limb, but was felt perceptibly in the heart. Ten days after the first treatment was given he walked without his crutches, and in six weeks resumed his usual vocation, entirely cured.

Another gentleman of the same city, whose right limb was double its natural size from this disease, was cured in five days.

Founded as it is upon common sense and scientific principles, this system of treatment, if properly administered, is absolutely infallible.

RHEUMATISM IN THE ARMS.

In this trouble we find the muscles of the shoulder at fault, obstructing either the brachial, axillary, or subclavian veins.

Treatment.

1st. Raise the arm as high and strongly as possible, but slowly, above the head. With the thumb of the disen-

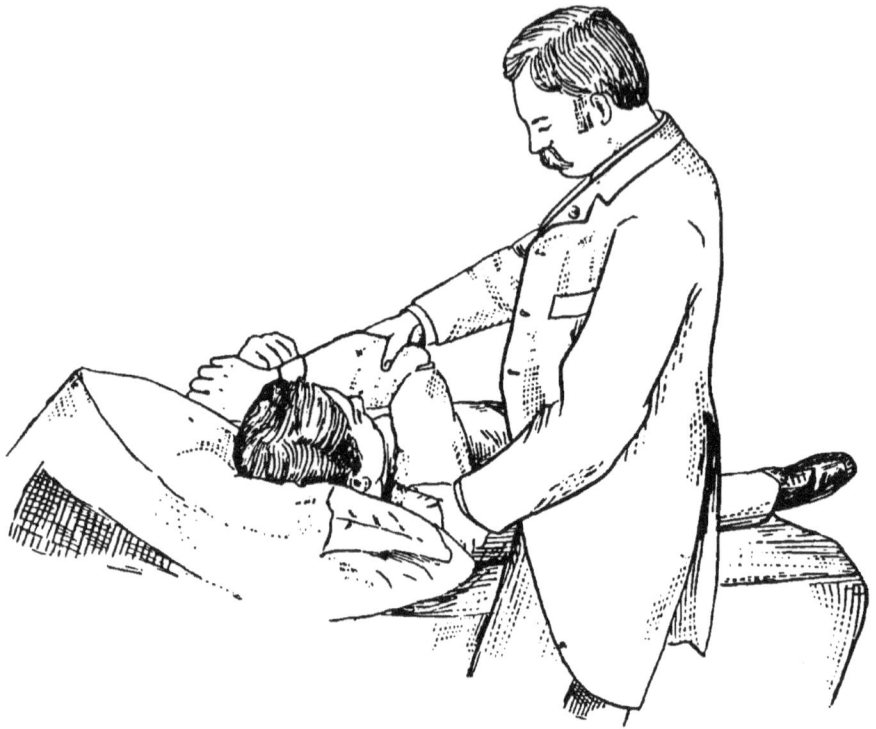

Cut 17.

gaged hand (beginning at the upper border of the scapula) press upward on the muscles between the scapula and spine, while raising the arm (see cut 14). Lower the arm with a backward motion, move the thumb down an inch, and again raise the arm, repeating until the lower border of the scapula is reached.

2d. Place one hand on the shoulder, pushing the muscles toward its point; with the other grasp the patient's elbow, and, while pressing hard with both hands, move the arm forward and upward around the head (cut 17).

3d. Grasp the arm with one hand close to the shoulder; with the other hold the arm from turning and move the muscles from right to left and from left to right the entire length of the arm.

4th. Stretch the arm, pulling slowly but very strongly.

This treatment not only cures any form of rheumatism, but paralysis and various forms of blood disorders in that member.

RHEUMATISM OF THE ENTIRE SYSTEM.

When inflammatory rheumatism extends over the entire system, the spinal column as well as the extremities should be treated.

Treatment.

1st. Place the patient on the side, and, using the arm as a lever (see cut 3), beginning with the fingers at the base of the neck, close to the spine, move the muscles upward and outward the entire length of the spinal column.

2d. Place one hand under the chin, the other under the back of the head; have an assistant take the feet, and pull steadily as hard as the patient can stand.

It is indeed surprising how quickly a patient responds to this treatment.

SCIATIC RHEUMATISM.

Sciatic rheumatism may be caused either by a pressure on the nerve itself in or near the thigh, or in the spine at the origin of the nerves that form the sacral plexus, of which the great sacro-sciatic nerve is a continuation.

Treatment.

1st. Flex the leg (with one hand grasping the ankle, the other resting on the knee) as far as possible toward the chest, slowly but strongly (cut 6).

2d. Extend the leg, turning the knee in, the foot out (cut 7).

3d. With one hand on each side of the thigh, move all the muscles from right to left and *vice versa*, very deep (cut 13).

4th. Place the patient on the side, and, beginning at the last dorsal vertebra, with the fingers close to the spine, move the muscles upward and outward from the spine to the end of the sacrum.

This treatment will cure the most stubborn cases of sciatic rheumatism in from two to six weeks.

LAME BACK.

Lame back, which may be traced to many different causes, can be invariably cured by our method; acute cases instantly, and chronic cases of many years' standing by a continuation of the treatment.

Treatment.

1st. Place the patient on the side, and, using the limb

Cut 18.

as a lever (cut 18), with the fingers close to the spine, commencing a little above the last lumbar vertebra (small of
the back), move the muscles up and out from the spine with
each upward motion of the limb.

2d. Extend the limb, move the hands down one inch,
and repeat until the lower part of the sacrum is reached.

3d. Place the patient on the back, and, grasping the
ankle, flex one limb after the other as far as possible toward
the chest, thus stretching the muscles of the back (cut 6).

4th. Place the patient on the face, and, with thumbs
on each side of the spine, beginning at the second lumbar
vertebra, press rather hard, moving the muscles upward;
move the thumbs down one inch, and repeat until you have
reached the second sacral vertebra; being very careful to
work thoroughly and deeply on each side of and a little below
the last lumbar vertebra (cut 19), as it is here the trouble
is usually found.

Here also is found the seat of kidney disease and female·
troubles. It is noticeable in those cases that the patient
usually has a weak back, and it has been demonstrated
beyond the shadow of a doubt that, working on these principles, not only the back, but the accompanying disorders, can
be entirely cured.

Of the many cases of lame backs treated by us, we will
mention that of a gentleman of Baxter Springs, Kansas. He
was assisted into our office, and told how, ten years before,
he was afflicted in a similar manner, being bedfast for six
months and on crutches two years. We gave him a treatment, not occupying over two minutes and curing him
instantly.

We might also mention the case of a gentleman of
Neutral, Kansas, cured by us in two weeks, after having
been given up by the medical fraternity.

ECZEMA.

Symptoms.

Inflammation of the skin, more or less redness, and closely packed vesicles, not larger than a pinhead, which run together, burst, and exude a starch-like fluid. This disease usually appears on the scalp, behind the ears, on the face, forearm, or the legs.

Eczema is caused by a contraction that holds the venous blood in the capillaries of any given part, thus causing a disease of the skin. It can be readily cured by kneading the flesh and stretching the muscles between the eruptions and the heart.

Among the many cases of this trouble cured by us, we will mention a lady of Galena, Kansas. The eczema was rapidly advancing toward the body on the right limb, and all known methods had failed to give relief. By giving the treatment prescribed for inflammatory rheumatism, thus freeing the circulation of the blood, she was cured in one month.

VARICOSE VEINS.

Symptoms.

The affected veins are dilated, tortuous, knotted, of a dull leaden or purplish blue color, with much discoloration of the parts and some swelling of the limb; if a great many small cutaneous veins are alone affected, they present the appearance of a close network; the enlarged veins and local swelling diminish after taking the horizontal position.

Varicose veins are caused by a stoppage of the veins, usually by a pressure on the long saphenous or femoral vein, and can be readily cured by stretching the muscles of the

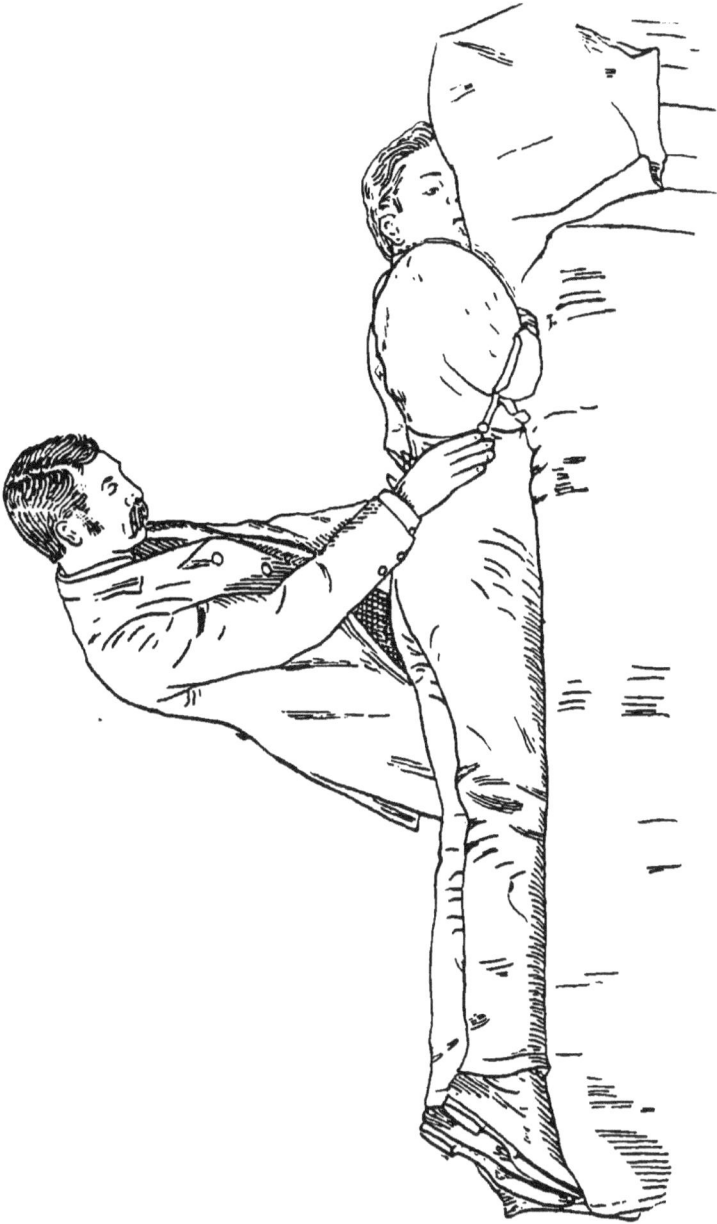

Cut 19.

thigh and otherwise treating the limbs as in inflammatory rheumatism, being very careful in handling the flesh below the knee.

We will mention our first case of varicose veins to prove to our readers how easily this disease may be cured by viewing the human system as a machine and the arteries and veins as rivers of blood, easily obstructed. The veins of the right limb below the knee were almost bursting, while the dead, stagnant blood in the capillaries formed sores, on which scales formed, occasionally dropping off, exposing the raw, bleeding surface beneath. Every known method had been tried and failed during the ten years he had suffered with this apparently incurable disease. The long saphenous vein, which empties into the femoral vein in the thigh, and whose branches gather and return the venous blood from the lower part of the leg, gorged, knotted, and distended as large as the little finger, could be traced to the obstruction, a contracted muscle in the thigh. It almost seems incredible that, where its cause was so apparent, for years the medical fraternity would work on the effect, encasing the limb in a rubber stocking to strengthen the bursting veins, while throwing medicine into the river above, with some object in view, unknown to the writer and possibly equally unknown to themselves. We gave the gentleman four treatments, stretching and freeing all the muscles of the thigh and starting the blood up the long saphenous vein. His limb immediately became easier. At this time we were called to southern Kansas, and after a year, when we had almost forgotten the incident, we met our old friend in a small Missouri town, entirely recovered.

THE THYROID GLAND.

The thyroid gland bears much resemblance in structure to other glandular organs, and was formerly classified together with the thymus, suprarenal capsules, and spleen, under the head of ductless glands, since, when fully developed, it has no excretory duct. The thyroid varies in weight from one to two ounces. It is larger in the female than in the male, and slightly increases in size during menstruation. It is situated at the upper part of the trachea, and consists of two lateral tubes, placed one on each side of that tube and connected by a narrow transverse portion, the isthmus. The arteries supplying the thyroid are the superior and inferior thyroid, and sometimes a branch from the innominate artery or arch of the aorta. The arteries are remarkable for their large size and frequent anastomosis; the veins form a plexus on the surface of the gland and on the front of the trachea, from which rise the superior, middle, and inferior thyroid veins. The two former terminate in the internal jugular, the latter in the innominate vein.

GOITRE.

Symptoms.

The thyroid gland is subject to enlargement, which is called goitre. For the relief of these growths various operations have been resorted to, such as the injection of tincture of iodine or perchloride of iron, ligature of the thymus, and extirpation of a part or the whole of the thyroid gland. The thyroid gland having an unusually large blood-supply, it is but reasonable to suppose that an obstruction to its veins would necessitate an enlargement of the gland, or goitre. It is equally reasonable that if the obstruction is removed, the

Cut 20.

goitre will soon be taken up by absorption and disappear. Acting on these principles, we have cured hundreds of goitres. It will be observed that the clavicle (collar-bone) is not nearly as prominent in people troubled with goitre as in those not afflicted with this disease. The contracting muscles and depressed clavicle, which are obstructing the escape of the blood from the thyroid gland, cause goitre. Raising the clavicle and stretching these muscles cures goitre, thus proving our theory to be correct.

Treatment.

1st. Stand behind the patient, and, extending the left arm around the neck, place the left thumb under the right clavicle (collar-bone) at about its middle; grasp the patient's right wrist with the disengaged hand, raise the arm slowly above the head, and lower with a backward motion, at the same time springing the clavicle up with the thumb of the left hand; raise the left clavicle in like manner (cut 20).

2d. Place the patient on the back, with one hand under the chin, the other under the back of the head; pull gently, rotating the head in any direction that will best stretch the muscles in the front and sides of the neck (cut 4).

3d. Place the fingers below the goitre, pulling it upward and kneading it gently.

This method will cure in from two to six weeks any goitre on which iodine has not been freely used. Treatment should be given every other day, and will not occupy over five minutes' time.

FLESHY TUMORS.

Fleshy tumors, like goitre, are caused by an obstruction to the veins draining any given part, and are easily cured by stretching and moving all the flesh and muscles in the immediate vicinity, and kneading and moving the tumor in all directions.

DISEASES OF THE HEAD.

Granulated eyelids, dripping eyes, inflammation of the eyes, catarrh, polypus of the nose, catarrhal deafness and roaring in the head, enlarged tonsils, mumps, erysipelas of the face, and many other diseases of the head, are caused by a contracted condition of the muscles of the neck, obstructing the flow of the venous blood on its return journey to the heart. The circulation being thus interrupted, disease at the weakest point is the result. To remove the cause by a general treatment of the neck is but the work of a moment, and never fails to effect a cure.

General Treatment of the Neck.

1st. Place the patient on the back, with one hand under the chin and the other under the back of the head; pull gently, rotating the head in all directions, slowly but strongly, endeavoring to stretch all muscles of the neck (see cut 4).

2d. Pull slowly on the head until the body moves.

3d. Draw the patient's arms slowly but very strongly high above the head.

GRANULATED EYELIDS.

Symptoms.

In this affection the conjunctival membrane, or white of the eye, is raised into little projections, presenting a rough, irregular appearance. It is a consequence of long-continued or maltreated inflammation, and if not cured, it may in time occasion opacities of the cornea by the irritation it causes, followed by blindness.

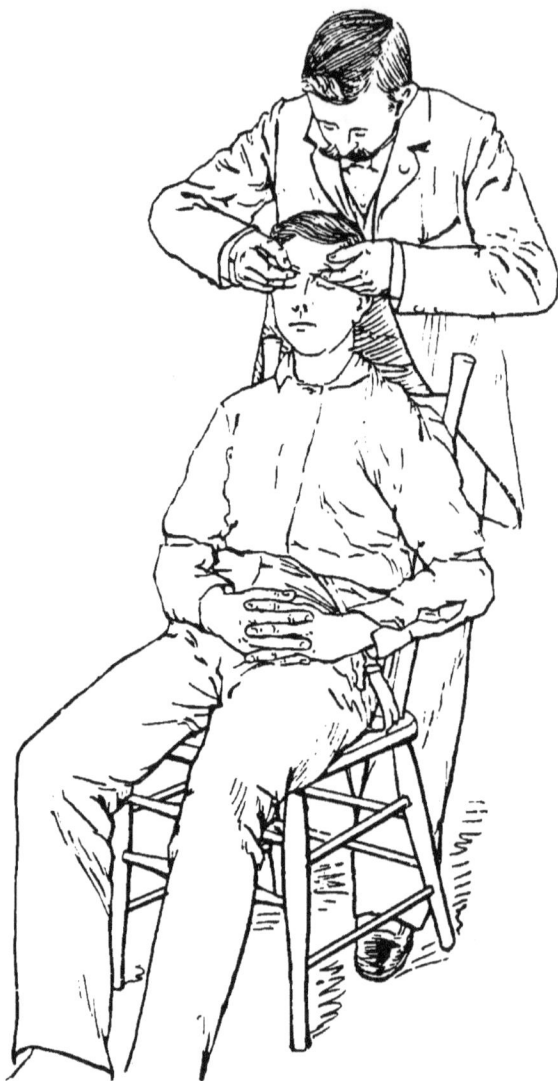

Cut 21.

Treatment.

1st. General treatment of the neck.

2d. Standing at the head of the table, with the index finger work gently, but as deeply as possible, moving the muscles and pressing under the edge of the bone entirely around the eye (cut 21); this frees the circulation and gives immediate relief.

3d. Pinch the eyelid gently wherever granules are formed, thus starting a natural circulation.

We have never known this method to fail, even in the worst cases. One old gentleman at Lewiston, Missouri, was cured by me in this manner, after suffering fifty years, in two months' time. The usual time required for an ordinary case is from two to four weeks. Treatment should be given every other day.

INFLAMMATION OF THE EYES.

Symptoms.

Inflammation of the eyes often comes on with a sensation as if sand had got into the eyes. In some instances this complaint proceeds no farther, but gradually goes; but at other times it is followed by heat, redness, and prickling, with darting pains.

Treatment.

1st. General treatment of the neck.

2d. With the index finger work gently but firmly under the edge of the bone surrounding the eye, thus freeing all ducts and glands, and also starting the circulation (cut 21).

3d. Place the fingers on the patient's temples, and, with a circular motion, move all the muscles as deeply as possible.

Treatment should be given once each day. This will

cure the most aggravated case of inflammation of the eyes in a few days.

DRIPPING EYES.

Dripping eyes are usually accompanied with a catarrhal difficulty in the lachrymal duct, which conveys all watery substances from the eye to the interior of the nose. Its obstruction causes the overflow at the eye; we must therefore cure the catarrh, and the eyes will take care of themselves.

Treatment.

1st. Give the general treatment for the neck.

2d. Free all the muscles around the eyes as in inflammation and granulation of the eyes (cut 21).

3d. Beginning deep in the corner of the eye, with the thumb on one side of the nose, the index finger on the other, move the flesh and muscles upward and downward its entire length.

This treatment is infallible, cases of twenty years' standing having been cured by us in one month's time. Treatment should be given every other day.

CATARRH.

Symptoms.

This disease sometimes prevails epidemically, and it is to this form medical writers apply the term "influenza," while cases that occur incidentally are called catarrh, or cold. When it prevails epidemically, it undoubtedly depends upon the state of the atmosphere, though in some cases it has been attributed to contagion.

In general, it comes on with a dull pain or sense of weight

in the forehead, sometimes preceded by a slight chill, redness of the eyes, and fullness and heat in the nostrils, which is soon followed by a discharge of thin acrid fluid from the nose, together with soreness in the windpipe, hoarseness, frequent sneezing, dry cough, loss of apppetite, and general lassitude; towards evening the pulse becomes considerably quickened and a slight fever arises.

Treatment.

1st.   General treatment of the neck.

2d.   Beginning deep in the corner of the eyes, move the muscles upward and downward, with as hard a pressure as patient can stand, the entire length of the nose.

3d.   Placing one hand on the back of the head, the other on the forehead, press very hard, moving the muscles of the forehead in all directions, especially those immediately over the eyes.

Treat each day for one week, when the patient will be on a fair way to recovery; one month's treatment will cure the most stubborn case.

CATARRHAL DEAFNESS AND ROARING IN THE HEAD.

Catarrhal deafness can be always benefited, and usually cured, by the catarrh treatment, after which the patient should endeavor to breathe while holding the nose and mouth, thus springing the drums of the ear and starting the circulation.

POLYPUS IN THE NOSE.

Symptoms.

When the polypus is located in the nose, there is a nasal sound in the voice, the patient acquires the habit of keeping his mouth open to facilitate breathing, there is difficulty of swallowing liquids, the nose is enlarged externally on the affected side, and on looking up the nostril the polypus may be seen. In consequence of the stuffy symptoms which a polypus occasions, it may at first be mistaken for a cold in the head; but, on the nose being violently blown, the polypus descends and appears near the orifice, causing the obstruction to return, contrary to the usual result of such an operation.

Treatment.

General treatment of the neck and other catarrhal treatment once each day. A cure is guaranteed.

MUMPS.

Symptoms.

At first there is a feeling of stiffness and soreness on moving the jaw, and the child complains of discomfiture on eating; indeed, the pain caused by eating, or even drinking, is sometimes agonizing. The glands under the ear soon begin to swell, and they continue to be sore and painful, with more or less fever and headache, for about a week. There is little danger, although there are instances in which, from exposure to cold or from cold applications, the disease has been transmitted to the testicles in boys and to the breasts of girls.

Treatment.

General and very thorough treatment of the neck once each day. Instant relief and a cure are guaranteed.

ENLARGEMENT OF THE TONSILS.

Enlargement of the tonsils may be of two kinds:

1. The common abscess, occurring in inflammatory sore throat.

2. Chronic swelling, generally the consequence of previous inflammation of the gland in a scrofulous person. They often become so large as to impede both respiration and swallowing.

Treatment.

General treatment of the neck once each day. A cure is guaranteed.

ERYSIPELAS IN THE FACE.

Symptoms.

Erysipelas is known by a spreading, inflammatory redness of the skin, with considerable puffy swelling, tenderness, burning, painful tingling, and tensions. The color varies from a faint red to a dark red or purplish color, becoming white under pressure, but resuming its former color at the removal of the pressure. An attack is usually ushered in with shivering, languor, headache, nausea, bilious vomiting, and the ordinary symptoms of inflammatory fever, accompanied or followed by inflammation of the part affected. When erysipelas attacks the face, it nearly always commences at the side of the nose or near the angle of the eye.

Treatment.

General and very thorough treatment of the neck. Instant relief and a cure are guaranteed.

WHOOPING-COUGH.

Symptoms.

The cough is accompanied by a shrill, reiterated whoop; vomiting is also a frequent incident. It is contagious under certain circumstances, which are not well understood. The disease comes on with the usual symptoms of catarrh. The whoop or sonorous spasm is frequently violent, the face becoming turgid and purple from suffusion and the eyeballs swollen and prominent. The paroxysms at first recur several times during the day, are most violent toward evening, and least so during the night. After the disease has continued some time, they return only in the morning and evening; and at the end of the disease, in the evening only. The violence of the·disease varies from the slightest indisposition, without feverishness, to the severest spasmodic agitation, attended with high and dangerous fever. In duration it varies from one week to one year, the usual period ranging from three weeks to three months,

Treatment.

1st. General treatment of the neck.

2d. Place the patient on the back, and, one operator grasping each wrist, raise the hands high above the head, at the same time placing the fingers of the disengaged hand between the patient's shoulders, close to the spine, at the upper border of the scapula, and pressing hard as the arms are drawn slowly up; lower the arms with a backward motion; move the fingers one inch down the spine, and repeat until the lower border of the scapula is reached (cut 1).

Treat each day until the disease is cured.

PNEUMONIA.

Symptoms.

Pneumonia generally comes on insidiously, with rest-lessness and feverish disturbance, and sometimes has made great progress before the true character of the disease has been discovered. There is a deep-seated, dull pain beneath the breast-bone or shoulder-blade; a great feeling of illness; frequent short cough, with expectoration of viscid matter of a green, yellow, or pale color, sometimes tinged with blood, which forms such tenacious masses that inversion of the vessel containing them will not detach them. Profuse green expectoration is la serious symptom. The breathing is hurried and difficult, the skin hot, especially in the regions of the armpits and ribs; there is no moisture in the nostrils, and there exists great thirst. If the disease is unchecked, the face often exhibits patches of redness and lividity and the blood-vessels of the neck become swollen and turgid. The patient may sink either from exhaustion or obstruction of the lungs.

Treatment.

1st. General treatment of the neck.

2d. Place the knee between the shoulders and raise the arms slowly, gently, but very strongly, high above the head, lowering them with a backward motion (cut 2).

Treat every few hours as the case may seem to require. Instant relief will be the result of this treatment, and in a few days a complete cure. We have taken numerous cases in their last stages, and never failed to be rewarded with the most gratifying results.

NEURALGIA.

Symptoms.

This is a disease of the nervous system, and the symptoms consist of severe paroxysms of pain, of a purely nervous character. The pain is generally very severe and more or less darting, sometimes burning, tearing, aching, and beating. In some cases it causes the patient to start suddenly, and spasmodic twitchings of the muscles are not uncommon. Sometimes there is tenderness of the part on pressure, especially on slight pressure, while hard pressure affords partial relief.

This disease is not confined to any particular part of the body. When it is in the nerves of the jaws and teeth, it causes one of the most distressing and unendurable forms of toothache. The eyes, temples, heart, spine, and stomach are not unfrequently attacked, and it frequently shifts from one to the other.

Treatment.

Neuralgia of any part should be treated in a similar manner to rheumatism of that part. If in the face, a general treatment of the neck should be given (page 134). A general treatment is often beneficial.

This disease is very stubborn, and while we have cured a great many cases instantly, in others we have been entirely baffled.

SAINT VITUS'S DANCE.

This is a disease characterized by convulsive movements of the limbs, occasioning ludicrous gesticulations, and arising from involuntary action of the muscles. It has been wittily termed "insanity of the voluntary muscles." This disease is caused by a pressure at some point along the spinal column upon the cerebro-spinal cord, and can be either cured or greatly benefited by this method of treatment.

Treatment.

1st. Give general treatment of the spine (cut 3, page 33).

2d. Place one hand under the back of the head, the other under the chin, and pull slowly, the patient relaxing all his muscles until the body moves on the table.

3d. While pulling gently, rotate the head from side to side (cut 4).

4th. Treat the extremities as in rheumatism.

Treatment should be given every other day.

FISTULA.

Fistula usually commences with swelling near the rectum, attended with great pain, hardness, and acute inflammation; the tumor advances slowly to suppuration, and matter is formed.

In some cases, however, the disease proceeds till an opening is formed, with very little pain—so much so, that the patient is ignorant of the time when it formed; but more generally the pain is very severe, swelling great, and suppuration very extensive, and, in consequence of the pressure upon the neck of the bladder or urethra, there is a suppression of urine.

Fistula may be caused either by a fall or riding a spirited horse, bruising and possibly dislocating the coccyx, or "tail-bone," as it is usually called, causing a pressure on the veins which return the blood from these parts.

Treatment.

1st. Place the patient on the side, and, using the limb as a lever (cut 18), with the hands, beginning at the second

lumbar vertebra, move the muscles upward and outward
from the spine with each upward motion of the limb, working down with the hands as low as the coccyx, very
thoroughly.

2d.   Flex the limbs strongly against the chest.

3d.   Pass the index finger up the rectum and move the
coccyx a little each treatment to its normal position.

This treatment will cure the worst case of fistula in from
two to six weeks.

PILES.

A sensation of heat, fullness, and perhaps itching, is
felt about the anus; the swelling increases until small tumors
form, which are sore and painful; these may be external and
visible or internal, and are often of a bluish color, and, when
inflamed, they are very sore and painful to the touch.   There
is frequently a discharge of blood, especially from internal
piles, and such discharges often return repeatedly until a
habit is established, and there is a feeling of fullness before
and relief after such discharges.

Piles that do not bleed are called *blind;* this variety is
apt to take on inflammation, when they become full, appear
ready to burst, and are so very sensitive the patient can
neither sit, lie down, nor walk.

Piles are really a varicose condition of the rectum, and
are usually the result of an obstruction of the hemorrhoidal
veins.

Treatment.

1st.   Piles are often caused by constipation, and in such
cases our constipation treatment (page 69) will usually effect
a cure.

2d. Place the patient on the face, and, with a thumb on each side of the spine, beginning at the first sacral vertebra, move the muscles very deeply upward and outward from the spine, working down to the end of the coccyx.

3d. If the patient has itching or bleeding piles, pass the index finger its entire length up the rectum, very carefully moving the inner muscles from side to side, thus freeing the circulation. In protuding piles they should be replaced, and the same internal treatment given.

There is no danger in this treatment, and we have never known it to fail to effect a cure in from two to six weeks. The internal treatment, which is rather painful, should be given but once a week, and always after flushing the bowels. Most cases of piles can be cured simply by our constipation treatment which removes the usual cause of this distressing complaint.

Of the many aggravated cases cured by us, we will mention the case of a gentleman of Scammon, Kansas, whose piles protruded an inch and had not been replaced for twenty years. We took the case as an experiment, hardly hoping to effect a cure. In three weeks our patient was entirely well, and up to the present time has had no return of his old trouble.

ABSCESSES.

Symptoms.

Abscesses first appear as a hot, hard swelling, accompanied by a burning, dull, throbbing sensation; as the swelling gradually increases the skin covering it assumes a purple or brownish red tint, and in a few days softens and suppuration takes place at several points; the matter is a thin, watery, and scanty discharge, but, if pressure be applied, a thick, glutinous fluid may be squeezed out.

Treatment.

1st. Move the flesh and muscles very deeply in all directions for some distance around the abscess.

2d. Move the abscess from side to side, pressing and rubbing it gently.

Any abscess can be cured in this manner in a very few days, and the pain instantly relieved.

COLD, HOT, OR ACHING FEET AND MILK LEG.

These troubles are caused by a contraction of the muscles, usually in the thigh in cold extremities, obstructing the free flow of arterial blood, while in the latter troubles it is the returning blood which is obstructed; in either case the heart, pumping the blood steadily against the obstruction, is at a disadvantage, and soon heart disease is the result. By using the leg as a lever, moving and stretching the muscles in all possible directions, we remove the cause, and an immediate cure is the result.

Treatment.

The same as in inflammatory rheumatism (page 117).

BACKACHE.

Treatment.

Place the patient on the side, and. beginning at the first lumbar vertebra, with the fingers close to the spine, with a steady pressure move the muscles upward and outward with a circular motion, moving the fingers down one inch after each upward motion until the lower part of the sacrum is reached; treat the other side in a similar manner.

Instant relief is always the result of this treatment, and, if continued a few days, a permanent cure. Treatment should be given each day.

BILIOUSNESS.

Symptoms.

There is more or less fullness and sensation of a load or other symptoms of uneasiness in the region of the stomach; there is languor, dull headache, or sleepiness, and sometimes slight yellowishness of the eyes and skin.

Treatment.

1st. General treatment of the spine (page 32).

2d. Place the knee between the shoulders on the fourth dorsal vertebra, and, while pressing hard, raise the arms slowly but very strongly high above the head, lowering them with a backward motion (cut 2).

The most stubborn case can be cured in this manner in from two to six weeks, treatment to be given every other day.

BALDNESS.

Baldness is caused by an obstruction in the cutaneous circulation, and can be prevented by our method of treatment.

Treatment.

1st. General treatment of the neck (page 134).

2d. Move the scalp in all directions, working it as loose as possible, once each day, thus freeing the blood-supply.

It is indeed surprising how soon this treatment will check the falling out of the hair and promote a renewed growth.

CANKER OF THE MOUTH.

Symptoms.

This disease is characterized by the membrane covering the sides of the tongue and inside of the cheeks becoming red and inflamed, and afterwards covered with large ulcers. The tongue becomes swollen, there is a profuse secretion of saliva, the breath is offensive, and swallowing is difficult.

Treatment.

1st. General treatment of the neck.

2d. Place the finger in the mouth and gently move the inflamed swollen surface as deeply as possible in all directions, rubbing the ulcers gently.

This treatment starts the circulation, gives immediate relief, and in a few days a complete cure. Treatment is to be given once each day.

CRICK IN THE NECK.

This is a form of rheumatism caused by a cold draught striking the neck. The muscles of that side of the neck contract, causing difficulty in moving the head. This can be instantly cured by a thorough general treatment of the neck.

EARACHE.

Symptoms.

An excessive throbbing pain in the ear. The more violent forms of this disease are attended with excruciating, throbbing pains, delirium, and sometimes convulsions.

Treatment.

1st. General treatment of the neck.

2d. Place the finger in the ear and move the muscles as deeply as possible.

3d. Place the lips close to the ear and blow gently, but very hard.

4th. Move all the muscles deeply immediately around the ear.

Instant relief is usually the result; in very stubborn cases several treatments may be necessary.

PIMPLES.

Pimples are caused by an obstruction to the cutaneous veins, and are easily cured by a general treatment of the neck (page 134).

DIZZINESS.

Symptoms.

The patient is suddenly seized with a sense of swimming in the head; everything appears to him to turn around; he staggers, and is in danger of falling. This complaint is attended with very little danger when it arises from hysterics or any nervous disorder, but when it arises from an unnatural quantity of blood in the head, there is danger of apoplexy. This complaint often proceeds from difficult or obstructed menstruation.

Treatment.

This disease can be cured in from two to six weeks by our general treatment (page 93), which will remove the cause and permit a natural flow of all the fluids of the body.

FITS AND CONVULSIONS.

While fits and convulsions can be instantly relieved by a genral treatment of the neck (page 134) and spine (page 32), as a rule, they are very stubborn, not over 50 per cent being cured by our method, and those cases only after long-continued treatment, to be given every other day.

HOARSENESS.

A general treatment of the neck once or twice each day will cure this trouble in a very short time.

CHRONIC GOUT.

This is a persistent constitutional affection, characterized by stiffness and swelling of various joints, with deposits of urate of soda.

Symptoms.

The deposits in the joints constitute the distinguishing feature. Chronic stiffness and swelling of various joints, with pain, are considered as cases of chronic rheumatism or gout.

Treatment.

1st. Treat the limbs as in inflammatory rheumatism every other day.

A continuation of this treatment will be of vast benefit to the patient, even though it may not effect a complete cure.

HIP DISEASE (WHITE SWELLING).

This is a very painful disease, and is usually seated on some of the joints of the body, principally the hip, knee, ankle, and elbow. The skin remains white, even in great inflammation. Sometimes the disease is rather mild in its character, at others very painful; the seat of the pain is in the periosteum, or covering of the bone, which in most cases becomes diseased and scales off.

Symptoms.

In the commencement there is a very severe pain felt deep in the joint, and when the person moves, the pain becomes intolerable; as it progresses, there is swelling, but no redness—a shining whiteness, with hardness or callous; it slowly increases until the swelling is very great; there is a discharge of matter; the limb wastes, becomes bent, and, when in the hip, osseous matter fills up the joint and slowly dislocates the head of the bone, either causing permanent dislocation or stiffness in the process of time; there are generally small pieces of bone detached; the patient is very thin, with much constitutional disturbance, hectic fever, etc.

Treatment.

This disease, if taken in a reasonable time, can be cured in a few treatments.

1st. Move the joint in all directions as strongly as the patient can stand.

2d. With the hands move all the flesh and muscles very deeply for some distance around the joint, thus freeing the circulation and permitting Nature to act.

After dislocation takes place, the treatment will loosen the joint and render the limb much more useful.

CRAMP IN THE LEGS.

Symptoms.

Sudden contraction of the muscles of the calf of the leg, frequently the result of indigestion.

Treatment.

1st. Flex the limb strongly against the chest (cut 6), extending it with a light jerk.

2d. Move all the muscles of the leg very deeply from right to left and left to right, beginning at the thigh.

This treatment will give instant relief, and a few treatments will effect a permanent cure.

COLIC.

Symptoms.

Severe twisting, griping pain in the abdomen, chiefly around the navel, relieved by pressure, so that the patient doubles himself up, lies on his stomach, or rolls on the floor, writhing in agony. The bowels are usually constipated, but there is a frequent desire to relieve them, although little is passed but wind; there is no fever, nor is the pulse even quickened unless after a time it becomes so from anxiety. The paroxysms of pain are owing to the efforts of the bowels above to force downwards the mass of accumulated gas or feces, while the lower portion is contracted. Colic is sometimes mistaken for inflammation of the bowels, but may be distinguished by the fact that in colic there is no fever, and in inflammation the fever is high and there is great acceleration of the pulse.

Treatment.

1st. Place the knee in the back at the twelfth dorsal vertebra, and raise the arms high above the head, lifting strongly while pressing hard with the knee.

2d. Bend the patient backward over the knee.

3d. Place patient on the back and knead the bowels as in constipation (page 70).

This treatment will usually give instant relief.

FAINTING.

Fainting is a loss of volition and muscular power with complete or partial loss of consciousness, due to defective nervous power. It has various causes: debility from constitutional tendencies, or from loss of blood or other animal fluids; emotional disturbances, fright, sudden joy or grief, hysteria, etc. Many persons faint on seeing blood or a wound.

Treatment.

1st. General treatment of the neck (page 134).
2d. General treatment of the spine (page 32).

THE UTERUS.

That our readers may gain a more correct understanding of our method of treating diseases peculiar to women, it will be necessary to refer once more to the anatomy.

The uterus is the organ of gestation, receiving the fecundated ovum in its cavity, retaining and supporting it during the development of the fœtus, and becoming the principal agent in its expulsion at the time of parturition. In the virgin state it is pear-shaped, flattened from before backward, and situated in the cavity of the pelvis between the bladder and rectum; it is retained in its position by the round and broad ligament on each side and projecting into the vagina below. Its upper end, or base, is directed upward and forward; its lower end, or apex, downward and backward in line of the axis of the inlet of the pelvis. The uterus measures about three inches in length, two in breadth at its upper part, and nearly an inch in thickness, and weighs from an ounce to an ounce and one-half.

The size, weight, and location of the uterus varies at

different periods of life and under different circumstances. In the fœtus the uterus is contained in the abdominal cavity, projecting beyond the brim of the pelvis. At puberty the uterus is pyriform in shape, and weighs from eight to ten drams; it has descended into the pelvis, the fundus being just below the level of the brim of this cavity. During menstruation the organ is enlarged and more vascular, its surfaces rounder, and the lining membrane of the body thicker, softer, and of a darker color. During pregnancy the uterus becomes enormously enlarged, and in the ninth month reaches the epigastric region. After parturition the uterus regains nearly its natural position and size, weighing about an ounce and a half.

THE FALLOPIAN TUBES.

The Fallopian tubes, or oviducts, convey the ova from the ovaries to the cavity of the uterus; they are two in number, one on each side, situated in the upper margin of the broad ligament, extending from each superior angle of the uterus to the sides of the pelvis; each tube is about four inches in length. The general direction of the Fallopian tubes is outward, backward, and downward. The uterine opening is minute, and will only admit a fine bristle; the abdominal opening is comparatively much larger.

THE OVARIES.

The ovaries are oval-shaped bodies, flattened from above downward, situated one on each side of the uterus in the posterior part of the broad ligament, behind and below the Fallopian tubes. Each ovary is connected by its anterior straight margin to the broad ligament, by its inner extremity to the uterus by a proper ligament, the ligament of the ovary, and by its outer end to the fimbriated extremity of the Fallopian tube. The ovaries are each about an inch and a half in length, three-quarters of an inch in width, and about a third of an inch in thickness. The uterus being suspended by muscles and ligaments in the cavity of the pelvis, and being subject to so many and such radical changes, it is in no way surprising that ulceration, polypus, cancer, and prolapse of the uterus are so very prevalent.

DISPLACEMENTS OF THE UTERUS.

The true pathology or proximate condition of these affections is but little understood by the medical profession, as is apparent from the general ill success attending the ordinary treatment. The term *prolapsus* is used indiscriminately for all degrees of simple descent or *falling of the womb;* but some books use the term *relaxation* when the descent is only to the middle of the vagina, *procidentia* when the uterus descends to the labia, and *prolapsus* when it protrudes externally. *Retroversion* is that form of displacement in which the fundus uteri descends toward the sacrum, the os uteri, or mouth of the womb, inclining towards the pubes. *Anteversion* is the reverse of the preceding, the fundus falling forward and the os uteri inclining backward. In *inversion* the organ is turned inside out while in a state of prolapse. In some cases the upper part of the vagina protrudes into the lower, constituting what is called prolapse of the vagina.

Symptoms.

Prolapse of the uterus is attended with a heavy, disagreeable, or painful dragging-down sensation at the lower part of the abdomen, aching or weakness about the small of the back, and, when severe, great difficulty or inability in walking. At first there is increased mucous secretion,

which increases by degrees until it acquires the character
of an obstinate leucorrhea.

When the uterus is *retroverted*, the bowels are irregular
or constipated, and from the pressure of the displaced organ
in the rectum behind and urethra in front there is more or
less difficulty experienced in expelling the contents of the
bowels and bladder. In this situation the womb often becomes congested, inflammatory, and enlarged, and every
attempt at walking is exceedingly painful and exhausting.
In bad cases the patient can only endure a fixed, motionless
position in her chair or bed. There is, too, usually considerable tenderness and tension of the whole abdomen.

Anteversion is a less frequent occurrence; it is denoted
by difficulty in walking, sense of weight or fullness in the
pelvis, with many of the symptoms of prolapse, and is attended with much less difficulty in evacuating urine and
fæces than retroversion.

Inversion is known by the organ hanging down externally; it is usually the result of violence in extracting the
placenta, but may occur from an adhesion of the placenta
or from polypous tumors. In some instances the falling of
the uterus or vagina drags along the bladder with it, constituting what is called complicated prolapse. In this case the
bladder, being deprived of the expulsory aid of the abdominal muscles, is incapable of evacuating its contents without
artificial assistance.

Genital excrescence consists in polypus or other tumors
issuing from the surface of the uterus or vagina; they are of
all sizes and various degrees of consistency, from the soft-
ness of the sponge to the firmness of leather.

Special Causes.

Although medical authors and professors of midwifery
are continually talking about "relaxation of ligaments,"

which hold the uterus in position, as the main cause of its
displacement, it is quite clear that this relaxation has nothing whatever to do with it; the yielding or elongation of the
ligament itself being an effect of the displacement. The natural supports of the uterus are the vagina and abdominal
muscles; if the former is greatly relaxed, the uterus will
descend, and the ligaments, being kept constantly on the
stretch, will finally elongate more or less; and if the abdominal muscles are greatly debilitated, they do not contract
vigorously, so as to keep up equable and uniform compression in all the various positions of the body, and hence the
uterus is liable to fall forward or backward, or to incline
laterally; and when both are badly relaxed and debilitated,
we find both conditions of displacement—falling down and
tipping transversely across the pelvis.

In corroboration of this view of the subject, we may
advert to the fact that all the cases of uterine displacement
met with in our practice, with the single and rare exception
of such as are produced by violence, occur in females who
suffer from the very circumstances which are most efficient
in inducing muscular relaxation of these parts, as constipation, piles, dyspepsia, nervous debility, mis-menstruation,
abortions or miscarriages, preternatural labors, etc.

It is a well-known fact that all cases of female troubles
are accompanied by a weak, lame back, and it is to this point
we trace the real cause of most cases of falling of the womb
and other troubles peculiar to women. Either by an accident or overwork, the muscles of the back, from the first
lumbar vertebra to the last sacral, have become strained,
causing contraction and a consequent pressure on the nerves
which control the organs of generation, thus breaking the
nervous current from the brain to these parts, interfering
with the circulation and permitting the muscles which hold

the organs of generation in place to relax. The fact that our treatment gives not only instant relief in most cases, but a permanent cure in all, if continued from three to six weeks, is ample proof that in female complaints, as well as in all other troubles to which it has been applied, the never-failing principles of Osteopathy are as superior to the old methods of healing as electricity is superior to the tallow candle

Treatment.

1st. If the patient is constipated, flush the bowels and give constipation treatment (page 69).

2d. Place the patient on the side; beginning at the first lumbar vertebra, with the fingers close to the spine, move the muscles gently but very deeply upward and outward from the spine, as low as the last sacral vertebra, being very thorough. In the sacral region move all the muscles upward very deeply, for about two inches, on each side of the spine, as it is here we strike foramina (openings in the bone) that transmit nerves directly to the organs in question.

3d. Insert the finger and move the womb gently to its normal position.

The local treatment is seldom if ever necessary. We have noticed, in our extensive practice, that while adjusting the uterus gave temporary relief, cases in which no local treatment was given recoverd as rapidly, thus proving that to free and stretch the muscles of the back, removing all pressure from the nerves, enabling them to regain control of the parts in question, would cause the muscles attached to the uterus to contract and draw that organ to its proper position. Immediately after the first treatment, the back will feel easier, and in a few weeks at most a complete and permanent cure will be effected. We take great pleasure in recommending this treatment to the public, it is so easily administered and so infallible.

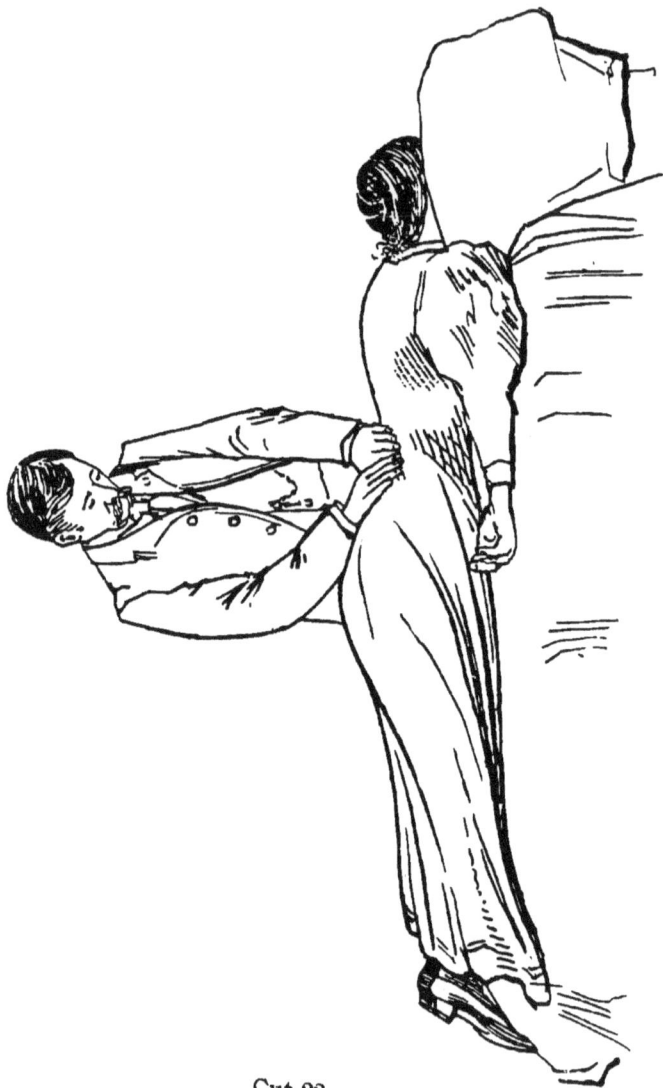

Cut 22.

SUPPRESSED MENSTRUATION.

Suppressed menstruation is attended with headache, difficult breathing, and palpitation; also languor and many dyspeptic symptoms, particularly a capricious appetite, and not infrequently a longing for innutrient and injurious sub-stances, as clay, slate-stone, charcoal, etc. In many cases there is an harassing cough and symptoms of a general decline.

Treatment.

1st.   Place the patient on the face, and, with the thumbs on each side of the spine, beginning at the first lumbar verte-bra, move the muscles very deeply upward and outward from the spine as low as the last sacral vertebra (cut 19).

2d.   Place the fingers close to the spine, and, with a steady pressure, draw the hands outward and upward on each side from the first lumbar vertebra to the sacrum (see cut 22).

3d.   Place one thumb on each side of the spine, begin-ning at the second lumbar vertebra, and, while pressing hard, have an assistant move the limbs gently to the left, raising them as high as the patient can bear and returning them to their former position, with a circular motion. Move the thumbs down one inch and repeat until the second sacral vertebra is reached (cut 23). Care should be taken to move the limbs slowly and to raise them only as high as the pa-tient can readily bear.

This treatment in these cases is infallible.   In ordinary cases two or three treatments will effect a cure; but one stubborn case in our experience, that of a young lady who had been sufferng for two years, took three months.

LEUCORRHEA.

Symptoms.

Generally there is a profuse mucous discharge from the utero-vaginal lining membrane of a white, cream, yellow, or greenish color, thin and watery or of the consistency of starch or gelatine, and it may be inodorous or fetid. When the discharge proceeds from the vagina, it is generally a light, creamy-looking fluid; in ulceration of the mouth of the womb it is profuse and semipurulent. In severe cases the whole system becomes injuriously affected; the face is pale or sallow, the functions of digestion are impaired, there are dull pains in the loins and abdomen, cold extremities, palpitation and difficult breathing after exertion, debility and loss of energy, and partial or entire suppression of the menstrual flow.

Treatment.

A thorough general treatment every other day, being very thorough in the lumbar region, will cure any case of this disease.

CHANGE OF LIFE.

Symptoms.

While the change is in progress there is commonly more or less functional disturbance of the general health, the nervous system especially manifesting various changes, such as vertigo, syncope, headache, flushes of heat, urinary troubles, pains in the back extending down the thighs with creeping sensations, heat in the lower part of the abdomen, occasional swelling of the lower extremities, itching of the private parts, mental irritability and restlessness, culminating seriously sometimes, especially in patients of a decided nerv-

Cut 23.

ous character. Sometimes menstruation ceases abruptly. The monthly period may be arrested by cold, fright, or some illness; earlier in life the suppression would have been followed by a return of menstruation after removal of the cause, but now Nature adopts this opportunity to terminate the function. Gradual termination is, however, more frequent and is attended with less disturbance of health. In gradual extinction one period is missed and then there is a return, a longer time elapses and there is an excessive flow; this continues for a time, the returns being fewer and farther apart, until they cease altogether.

At this critical period there is not infrequently an *enlargement of the abdomen*, which, though it may occur earlier in life, is due to causes peculiar to this.

Treatment.

A general treatment every other day will equalize the circulation and give wonderful relief (page 93).

GATHERED BREASTS.

Symptoms.

When inflammation occurs in the tissues behind the breast and on which it is placed, the pain is severe, throbbing, deep-seated, and increased by moving the arm and shoulder; the breast becomes swollen, red, and more prominent, being pushed forward by the abscess behind. Sometimes, but less frequently, the breast itself is involved, when the pain becomes very acute and cutting, the swelling very considerable, and there is much constitutional disturbance— quick, full pulse, hot skin, thirst, headache, sleeplessness, etc. This variety of gathered breasts is preceded by *rigors* (shivering fits), followed by heat.

Treatment.

1st. Raise the arms high above the head, with the knee between the shoulders, lowering the arms with a backward motion.

2d. Move all the muscles near the breasts very deeply.

3d. Move the breasts gently in all directions, raising them up and endeavoring to free all the glands, muscles, and circulation.

Treat every few hours. Immediate relief, and a cure in a very short time, will be the result.

OBSTETRICS.

In all cases of obstetrics, except those in which through some malformation it becomes necessary to use instruments in effecting the delivery, Osteopathy is indeed a grand success, diminishing the hours of labor at least three-fourths and reducing the suffering of the patient in a most remarkable manner.

Treatment.

1st. During the first stage of labor, with a finger on either side of the clitoris, press gently but rather hard; this pressure will cause a painless and complete dilatation of the cervix in a comparatively short time. Remove the fingers for an instant, and the patient screams with pain; resume the pressure, and instant relief is the result.

2d. In the second stage, as soon as the bearing-down pains begin, press gently but rather hard from the second to the last lumbar vertebra, on each side of the spine. As long as the pressure is continued, your patient will suffer no pain; remove your hands for an instant, and she cries: "Oh, Doctor! my back! my back!" It seems that a pressure at the points in question must, in some remarkable manner, break the nervous current between the brain and the muscles that are resisting the delivery of the fœtus, thus depriving them of their power of contraction and permitting the almost unobstructed and painless delivery of the child.

3d. During the interval of rest between the birth of the child and the delivery of the placenta flex the limb upon the chest, and, while an assistant presses hard on the great trochanter, extend the leg, abducting the knee and adducting the foot as much as possible; it will take but a moment and remove the cause of the numerous aches and pains in the hips and limbs, often the result of confinement.

If the physician or midwife in attendance will adopt this method as an accessory to their usual treatment, they will be surprised and gratified at the results.

THE SKELETON.

The entire skeleton in the adult consist of 200 bones.

The vertebræ are 33 in number, and are called cervical, dorsal, lumbar, sacral, and coccygeal, according to the position which they occupy; 7 being found in the cervical region, 12 in the dorsal, 5 in the lumbar, 5 in the sacral, and 4 in the coccygeal. The average length of the spine is about 2 feet and 2 or 3 inches; of this length the cervical part measures about 5, the dorsal 11, the lumbar about 7 inches, and the sacrum and coccyx the remainder.

The bones of the cranium are 8 in number, while those of the face number 14.

The os hyoides, sternum, and ribs, 26.

The upper extremities, 64.

The lower extremities, 62.

The bones of the upper extremities consist of the clavicle, or collar-bone; the scapula, or shoulder-blade; the humerus, the longest and largest bone of the upper extremities; the ulna, so called from its forming the elbow; the radius, lying side by side with the ulna; 8 carpal or wrist bones, 5 metacarpal or bones of the palm, and the 14 bones of the phalanges or fingers.

The bones of the lower extremities consist of the os innominatum, so called from its bearing no resemblance to any known object, which, with its fellow, forms the sides and anterior wall of the pelvis (in the young it consists of three separate bones, and although in the adult they have become united, it is usually described as the ileum, ischium, and os pubes); the femur, or thigh-bone, the largest, longest, and strongest bone in the body; the patella, in front of the knee-joint; the tibia, the fibula, 7 tarsal and 5 metatarsal bones, and 14 phalanges.

www.ingramcontent.com/pod-product-compliance
Lightning Source LLC
Chambersburg PA
CBHW021808190326
41518CB00007B/501